Oxford Chemistry Series

Oxford Chemistry Series

1972
1. K. A. McLauchlan: *Magnetic resonance*
2. J. Robbins: *Ions in solution* (2): *an introduction to electrochemistry*
3. R. J. Puddephatt: *The periodic table of the elements*
4. R. A. Jackson: *Mechanism: an introduction to the study of organic reactions*

1973
5. D. Whittaker: *Stereochemistry and mechanism*
6. G. Hughes: *Radiation chemistry*
7. G. Pass: *Ions in solution* (3): *inorganic properties*
8. E. B. Smith: *Basic chemical thermodynamics*
9. C. A. Coulson: *The shape and structure of molecules*
10. J. Wormald: *Diffraction methods*
11. J. Shorter: *Correlation analysis in organic chemistry: an introduction to linear free-energy relationships*
12. E. S. Stern (ed): *The chemist in industry* (1): *fine chemicals for polymers*
13. A. Earnshaw and T. J. Harrington: *The chemistry of the transition elements*

1974
14. W. J. Albery: *Electrode kinetics*
16. W. S. Fyfe: *Geochemistry*
17. E. S. Stern (ed): *The chemist in industry* (2): *human health and plant protection*
18. G. C. Bond: *Heterogeneous catalysis: principles and applications*
19. R. P. H. Gasser and W. G. Richards: *Entropy and energy levels*
20. D. J. Spedding: *Air pollution*
21. P. W. Atkins: *Quanta: a handbook of concepts*
22. M. J. Pilling: *Reaction kinetics*

G. C. BOND
PROFESSOR OF INDUSTRIAL CHEMISTRY, BRUNEL UNIVERSITY

Heterogeneous catalysis:
principles and applications

Clarendon Press · Oxford · 1974

Oxford University Press, Ely House, London W.1

GLASGOW NEW YORK TORONTO MELBOURNE WELLINGTON
CAPE TOWN IBADAN NAIROBI DAR ES SALAAM LUSAKA ADDIS ABABA
DELHI BOMBAY CALCUTTA MADRAS KARACHI LAHORE DACCA
KUALA LUMPUR SINGAPORE HONG KONG TOKYO

PAPERBACK ISBN 0 19 855412 5
CASEBOUND ISBN 0 19 855451 6

PRINTED IN GREAT BRITAIN BY
J. W. ARROWSMITH LTD., BRISTOL, ENGLAND

Editors' foreword

WHEN a chemical reaction which is favoured thermodynamically proceeds too slowly to be of use, the kinetics can often be made favourable by the adroit use of a catalyst. Modern chemical processes depend for their viability on a judicious choice of catalyst: but what are the criteria of selection? Why are some catalysts more effective than others? Can a catalyst be chosen to direct the course of reaction along a specific route, and to yield an intermediate rather than the thermodynamically more stable product? How does a catalyst function? How should it be used? These are some of the questions vital to a vigorous chemical industry, and their answers depend on a broad knowledge of chemical and economic principles. This book surveys the field of heterogeneous catalysis, and by highlighting a number of real industrial processes shows how one should think about the problems involved, and how they have been answered.

The basic thermodynamics and chemical kinetics used in this book are built on the topics discussed in Smith's *Basic chemical thermodynamics* (O.C.S. 8) and Pilling's *Reaction kinetics* (O.C.S. 22). The industrial applications of catalysis—the problems encountered, and the chemical and economic choices that have to be made—are described in *The chemist in industry* sequence of this series (edited by Stern): see (1) *Fine chemicals for polymers* (O.C.S. 12) and (2) *Human health and plant protection* (O.C.S. 17). The fundamental process of transferring an electron between a solid metal and a solution is described in detail by Albery in his *Electrode kinetics* (O.C.S. 14).

P.W.A.
A.K.H.

Preface

AN understanding of the phenomenon of catalysis requires some familiarity with all three classical branches of chemistry. Catalysts are most often inorganic materials, and a knowledge of inorganic chemistry is needed for their preparation. The reactions to be catalysed are frequently organic ones, and so an appreciation of the reactivity of organic molecules is helpful. The reactions themselves are studied by the techniques of physical chemistry. Furthermore the large-scale practice of catalysis additionally necessitates the application of chemical engineering principles, and a deep insight into the properties of materials. Studying catalysis is therefore an academically valuable exercise for students, since it demonstrates clearly the essential unity of scientific and technological problems.

This book is intended to give a survey of heterogeneous catalysis, and of its more important industrial applications, to a level adequate for Honours Degree students in British universities and polytechnics. It divides into two parts: the first four chapters describe the basic principles, while the following seven shorter chapters deal with applications. I have paid especial attention to the physical forms of catalysts used in industry, and to the types of reactor in which they are employed. This information is not usually to be found in chemistry textbooks. I have also included a short treatment of how catalysts are used in control of environmental pollution, which is a rapidly growing area of application.

I have attempted to dispel some of the aura of magic with which catalysis is often surrounded in students' minds. I have tried to explain how catalysis must be regarded as an essentially chemical phenomenon, and how it operates within, and indeed because of, well-established chemical principles. The chemistry which goes on at surfaces is no less important and even more intriguing than that which proceeds in homogeneous phases. Indeed I might have called this book 'Chemistry in two dimensions', for that is what catalysis is: but, had I done so, I might have failed in my chief objective in writing it, which was to share with as many people as possible the enjoyment I have had in studying this useful and entertaining branch of chemistry.

I am indebted to Dr. P. B. Wells of Hull University, who read and commented helpfully on the first four chapters; to Professor J. W. E. Coenen of the Catholic University, Nijmegen, who provided me with information on the technical aspects of fat hardening; to my secretary, Mrs. Gill Lewis, for her immaculate typing of my manuscript; and to my wife for her unfailing support.

G. C. BOND

Acknowledgment
I am grateful to Professor R. M. Barrer for permission to reproduce the illustrations for Figs. 2.13, 2.14, and 2.15, which appeared in *Chemistry and Industry*, pp. 1206–1207 (1968).

Contents

1. Basic Principles of Catalysis

1.1. Requirements for an industrially useful chemical reaction

IN the study of a reacting chemical system, two considerations are of importance. The first is, will the reaction proceed, and if so how far? All reactions must stop short of absolute completion: at what particular equilibrium position will the system come to rest? The answers to these and related questions are the province of chemical thermodynamics. The second consideration is, how fast does the reaction go? How quickly is the equilibrium situation attained? The answers to these and cognate questions are the concern of chemical kinetics. In the design of a chemical process, both considerations matter. The thermodynamics of the system determines the maximum attainable yield of products under specified conditions. No matter how fast the rate, if the yield is low, if the equilibrium constant is small, a viable process cannot result. On the other hand, it is of no value to have a system with a large equilibrium constant, where a high yield of product is potentially attainable, if that equilibrium can only be attained very slowly. In almost every case, economic factors require both favourable yields and rates.

Chemical thermodynamics is a well-established branch of science. Much information is available on very many chemical systems, from which the position of equilibrium may be estimated as a function of temperature, pressures or concentrations, and other variables. Even when accurate information is lacking, the corpus of thermodynamic knowledge may be used to make an intelligent guess at the answer; it constitutes our chemical intuition, and the use of approximate estimates can be of great value. A very brief outline of the relevant thermodynamic principles will now be given. For further information see E. B. Smith: *Basic chemical thermodynamics* (OCS 8).

1.2. Chemical thermodynamics

Let us consider the system

$$a\mathrm{A} + b\mathrm{B} \rightleftharpoons c\mathrm{C} + d\mathrm{D}.$$

The state of equilibrium is defined by the condition that the rates of the forward and reverse reactions are equal. Thus if the reactants and products are gases at low or moderate pressures we may write

$$r_\mathrm{f} = k_\mathrm{f} P_\mathrm{A}^a P_\mathrm{B}^b = r_\mathrm{r} = k_\mathrm{r} P_\mathrm{C}^c P_\mathrm{D}^d,$$

where r_f and r_r are the rates, and k_f and k_r the rate coefficients of the forward and reverse reactions respectively. Hence

$$\frac{k_\mathrm{f}}{k_\mathrm{r}} = \frac{P_\mathrm{C}^c P_\mathrm{D}^d}{P_\mathrm{A}^a P_\mathrm{B}^b} = K_P,$$

where K_P is the *equilibrium constant*. The above relation is often described as the Equilibrium Law, or Law of Mass Action. It signifies that for every chemical system of opposing reactions there is one and only one value of the equilibrium constant at a given temperature: this value depends only on the thermodynamic properties of the reactants and products. The most important function, for the constant pressure systems with which most of us work, is the Gibbs free energy G. The sign of the change in G accompanying a reaction determines whether or not it is thermodynamically feasible; for the reaction to occur spontaneously, ΔG must be negative. The magnitude of $-\Delta G$ determines how far the reaction will go: the more negative the value of the change in G, the larger will be the value of K_P. The relation between the two is expressed quantitatively by the *reaction isotherm*

$$-\Delta G^{\ominus} = RT \ln K_P,$$

the superscript signifying that this is the *standard* free energy change, i.e. that which results when reactants at unit pressure are transformed to products at unit pressure.

It is interesting to see how the equilibrium yield of product varies with the initial pressure of one of the reactants. Figure 1.1 shows how in the case of the simple equilibrium

$$\mathrm{A} + \mathrm{B} \rightleftharpoons \mathrm{C}$$

the final pressures of B and C alter as the initial pressure of A is increased. The initial pressure of B and the equilibrium constant are both taken as unity in the appropriate units. The calculations are very straightforward; the student might like to show for himself how the picture changes with the assumed value of K. We shall encounter curves of similar form at a later stage and in another context.

Differentiation of the reaction isotherm with respect to temperature leads to the *reaction isochore*:

$$\frac{\mathrm{d} \ln K_P}{\mathrm{d}T} = \frac{\Delta H^{\ominus}}{RT^2},$$

where $\Delta H^{\ominus}$ is the standard enthalpy (or heat content) change at constant pressure: this is thus readily determined from the temperature dependence of K_P. One further thermodynamic property of great importance is *entropy*, given the symbol S. The change in free energy and in enthalpy for an isothermal process are related by the equation:

$$\Delta G = \Delta H - T\,\Delta S.$$

Smith (1973) or standard textbooks of physical chemistry must be consulted for further information on the thermodynamics of chemical systems.

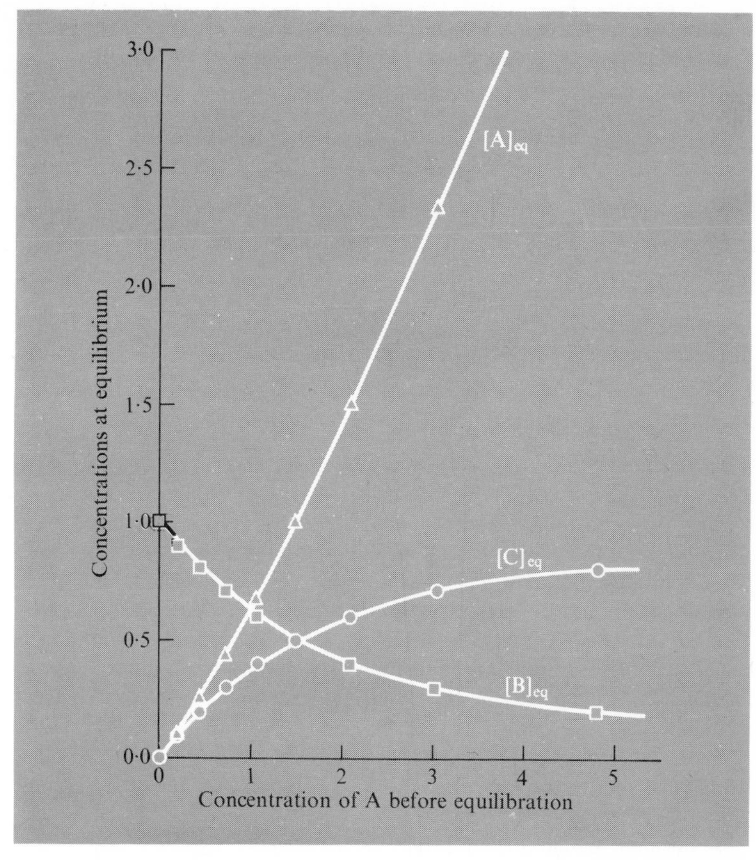

FIG. 1.1. The chemical isotherm: variation of equilibrium concentrations with initial concentration of A, with $K = 1$. In practice the concentration of a gas is measured by its pressure.

1.3. Chemical kinetics

Thermodynamic quantities are unfortunately quite useless for predicting the rate of a chemical transformation. Thermodynamics only treats energy differences between initial and final states: the rate of a reaction is decided by the energy barrier which reactants must surmount in the course of their transformation to products, and the height of this barrier is determined by the mechanism of the reaction. We must be careful too not to equate a, b, c, and d in the above relations with the experimentally determined orders of reaction. In many reactions there is a complex sequence of elementary steps intervening between reactants and products; one of these is usually slower than the rest and hence will be rate-limiting. The manner in which reactant pressures influence the rate of this process are expressed in the following way:

$$r = kP_{\mathrm{A}}^{m}P_{\mathrm{B}}^{n}$$

where m and n are the orders of reaction in A and B respectively, and k is the rate coefficient. This is numerically equal to the rate when the reactants are at unit pressure, but the units of k are dependent on $m+n$. This point is treated further in M. J. Pilling's *Reaction kinetics* (OCS 22), and in all physical chemistry textbooks. The orders of reaction give some indication as to how the reactants enter the rate-controlling step: but the equilibrium rates r_{f} and r_{r} are influenced by the fast steps as well as the slow step, so that a, b etc. are not equal to the orders of reaction.

For most chemical reactions, the rate coefficient k varies with absolute temperature T according to the Arrhenius equation:

$$k = A\exp(-E/RT),$$

where A is the so-called pre-exponential factor, E is the activation energy and R is the gas constant. The activation energy measures the amount of energy which reactant molecules must have before they can overcome the barrier between them and the product state. In a simple homogeneous gas-phase reaction, the exponential term is related to the Boltzmann distribution of kinetic energy and specifies the fraction of molecules having kinetic energy greater than E. The *collision theory* then suggests that the velocity constant should be given by

$$k = Z_{\mathrm{AB}}\exp(-E/RT)$$

where Z_{AB} is the collision number between unlike molecules: but, even in the case of simple bimolecular reactions, rates calculated by this expression are frequently much greater (and occasionally less) than those observed, and this has led to the idea that there is a steric as well as an energetic requirement. The introduction of the empirical steric factor P gives

$$k = PZ_{\mathrm{AB}}\exp(-E/RT)$$

but this is of no great assistance in the *a priori* calculation of reaction rates. In particular it is difficult to see how the collision theory can meaningfully be applied to reactions between molecules adsorbed on surfaces.

Of greater potential utility is the *transition-state theory* or the *theory of absolute reaction rates* (see Pilling (1974)). This focuses attention on the state of the reactants at the top of the energy barrier: this is the so-called *transition state* or *activated complex* and if it is considered to be in equilibrium with the reactants, then it can be treated by thermodynamic techniques. The following is an outline of this theory.

Consider the bimolecular reaction

$$\mathrm{A} + \mathrm{B} \underset{k_2}{\overset{k_1}{\rightleftharpoons}} \mathrm{AB}^{\ddagger} \xrightarrow{k_3} \mathrm{C} + \mathrm{D},$$

where $AB^{\ddagger}$ represents the transition state, and suppose that it decomposes to products sufficiently slowly that the equilibrium with reactants is not significantly disturbed. We may therefore write

$$\frac{k_1}{k_2} = K_P^{\ddagger} = \frac{P_{\mathrm{AB}^{\ddagger}}}{P_{\mathrm{A}}P_{\mathrm{B}}} \quad \text{or} \quad P_{\mathrm{AB}^{\ddagger}} = K_P^{\ddagger}P_{\mathrm{A}}P_{\mathrm{B}}.$$

The rate of the forward reaction is then given by

$$r_{\mathrm{f}} = k_3 P_{\mathrm{AB}^{\ddagger}}.$$

The complex flies apart when a vibrational mode changes to a translational mode, since translation is an irreversible vibration. The frequency ν of this decomposition is thus

$$\nu = \varepsilon/h$$

where h is Planck's constant and ε the average energy of the relevant vibration. If this is a thoroughly excited mode at the temperature T, then

$$\varepsilon = \boldsymbol{k}T \text{ and } \nu = \boldsymbol{k}T/h.$$

Thus:

$$\begin{aligned} r_{\mathrm{f}} &= kP_{\mathrm{A}}P_{\mathrm{B}} \\ &= K_P^{\ddagger}P_{\mathrm{A}}P_{\mathrm{B}}\boldsymbol{k}T/h \end{aligned}$$

and

$$k = K_P^{\ddagger}\boldsymbol{k}T/h.$$

The term $\boldsymbol{k}T/h$ has the units of frequency, and is $6 \times 10^{12}\ \mathrm{s}^{-1}$ at 300 K.

The thermodynamic relationships given above may now be applied to the equilibrium constant $K_P^{\ddagger}$, and analogous thermodynamic quantities defined. The *free energy of activation* is given by

$$-\Delta G^{\ddagger} = RT \ln K_P^{\ddagger}$$

and furthermore

$$\Delta G^{\ddagger} = \Delta H^{\ddagger} - T\,\Delta S^{\ddagger}$$

where $\Delta H^{\ddagger}$ and $\Delta S^{\ddagger}$ are respectively the *standard enthalpy* and *standard entropy of activation*. Then:

$$\begin{aligned} k &= \frac{\boldsymbol{k}T}{h}\exp(-\Delta G^{\ddagger}/RT) \\ &= \frac{\boldsymbol{k}T}{h}\exp(\Delta S^{\ddagger}/R)\exp(-\Delta H^{\ddagger}/RT). \end{aligned}$$

Comparing this with the Arrhenius equation we see that $-\Delta H^{\ddagger}$ replaces the activation energy E and

$$A = \frac{kT}{h}\exp(\Delta S^{\ddagger}/R).$$

In fact E equals $-\Delta H^{\ddagger}$ provided there is no change in the number of molecules in the reaction. We shall make further use of these expressions subsequently: the interested reader is referred to M. J. Pilling's *Reaction kinetics* (OCS 22) and other books in the bibliography for a deeper discussion of this theory.

1.4. Definition of catalysis

The foregoing short outline of the thermodynamics of chemical systems and of the kinetics of chemical reactions is necessary before we can discuss the concepts of catalysis in proper perspective. This we may now begin to do. It has long been known that the rates of many chemical reactions can be affected by traces of alien material which may be adventitiously present in the system or may be added deliberately. The word 'alien' is used to imply that the material it describes does not appear in the stoichiometric equation for the reaction. Such a material is termed a *catalyst* and it is defined as *a substance which increases the rate at which a chemical reaction approaches equilibrium, without being consumed in the process.* The phenomenon occurring when a catalyst acts is termed *catalysis.* It was first used in the scientific era by Berzelius who in 1836 used it to describe a variety of diverse observations concerning the effects of trace substances on reaction rates. Examples quoted included the acid-catalysed hydrolysis of starch to glucose, the effect of metal ions on the decomposition of hydrogen peroxide and the effect of platinum on the reaction of hydrogen with oxygen. This last effect had been studied by Michael Faraday. The word catalysis comes from two Greek words, the prefix *cata-*, meaning down, and the verb *lysein*, meaning to split or break. It is closely related to the word *analysis*, which means of course breaking up for the purpose of estimating the constitution of a substance or mixture. Berzelius probably used 'catalysis' to denote the breaking down of the normal forces which inhibit the reactions of molecules. The same word was also used in ancient Greece to denote a failure of social or ethical restraints, and for example what we would describe as a riot was called by them a catalysis. The word 'catalysis' now frequently appears in the popular press, but usually in the sense of 'bringing together', which is far from its true meaning. In this context the Chinese words *tsoo mei* which are used for a catalyst, and which also mean 'marriage broker', perhaps more accurately reflect the layman's idea of catalysis.

The primary effect of a catalyst on a chemical reaction is, as stated in the above definition of catalysis, to increase its rate: this means therefore to increase its rate coefficient. The consequent effects may be analysed either in terms of the collision theory or the transition-state theory.

According to the collision theory, the rate coefficient k is given by

$$k = PZ \exp(-E/RT).$$

Let us consider for simplicity a unimolecular catalytic transformation in which the slow step is the adsorption of the reactant: the collision number in question is then the number of collisions per unit time between the reactant molecules and the catalytic sites or species. Now the concentration of the latter will be quite small and so the number of relevant collisions will be much smaller (by a factor of some 10^{12}) than the number of collisions between reactant molecules alone, which number is relevant to the uncatalysed reaction but irrelevant to the catalysed reaction. Therefore if the catalysed reaction is to compete effectively with the uncatalysed reaction, then its exponential term must be some 10^{12} times larger, which means that its activation energy must be about 65 kJ mol^{-1} less. There may be some small relief in the form of a higher steric factor but this is unlikely to contribute more than a factor of 10^2 or 10^3 at most, and the main conclusion is not really altered. Neglecting this effect, we see that an activation energy difference of 65 kJ mol^{-1} only makes the rates of the catalysed and uncatalysed reactions equal: this scarcely represents efficient catalysis, for which the activation energy difference typically must exceed 100 kJ mol^{-1}.

In terms of the transition-state theory, we have seen above that

$$k = \frac{\mathbf{k}T}{h} \exp(-\Delta G^{\ddagger}/RT)$$

and so the effect of a catalyst must be to decrease the free energy of activation of the reaction. This in turn is composed of an entropy and an enthalpy of activation. Now the entropy of activation in a catalysed reaction will usually be less than in the corresponding uncatalysed reaction because the transition state is immobilized on the catalyst surface with consequent loss of translational freedom. There must therefore be a corresponding decrease in the enthalpy of activation to compensate for this, or to overcompensate for efficient catalysis. Thus according to either theory the activation energy for a catalysed reaction ought to be less than for the same uncatalysed reaction. There is much experimental evidence to show that such is indeed the case: some values of activation energies for catalysed and uncatalysed reactions are given in Table 1.1.

This situation may be shown both as a potential-energy profile (Fig. 1.2) and as an Arrhenius diagram (Fig. 1.3): the latter makes it plain that the effect of a catalyst is either to increase the rate at a given temperature (except at very high temperatures) or to decrease the temperature at which the reaction achieves a given rate. It must be emphasized that the lowering of the activation energy is a fundamental principle of catalysis, and applies to all forms of catalysis—homogeneous, heterogeneous, and enzymatic.

TABLE 1.1

Activation energies (kJ *mol*$^{-1}$) *for uncatalysed and catalysed reactions*

Reaction	E (uncatalysed)	E (catalysed)	Catalyst
$2HI \rightarrow H_2 + I_2$	184	—	—
	—	105	Au
	—	59	Pt
$2N_2O \rightarrow 2N_2 + O_2$	245	—	—
	—	121	Au
	—	134	Pt
Pyrolysis of $(C_2H_5)_2O$	224	—	—
	—	144	I_2 vapour

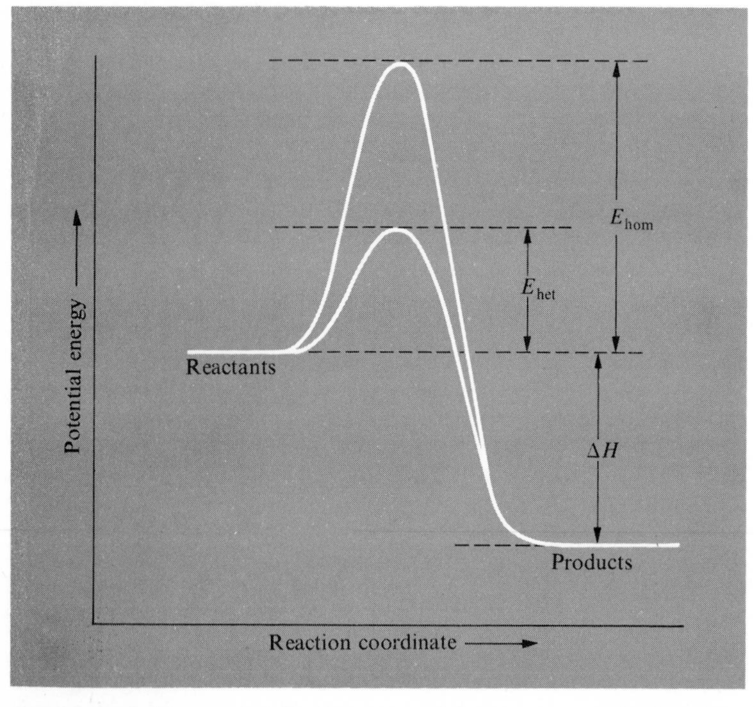

FIG. 1.2. Potential-energy profile for an exothermic reaction, showing the lower activation energy of the catalysed reaction.

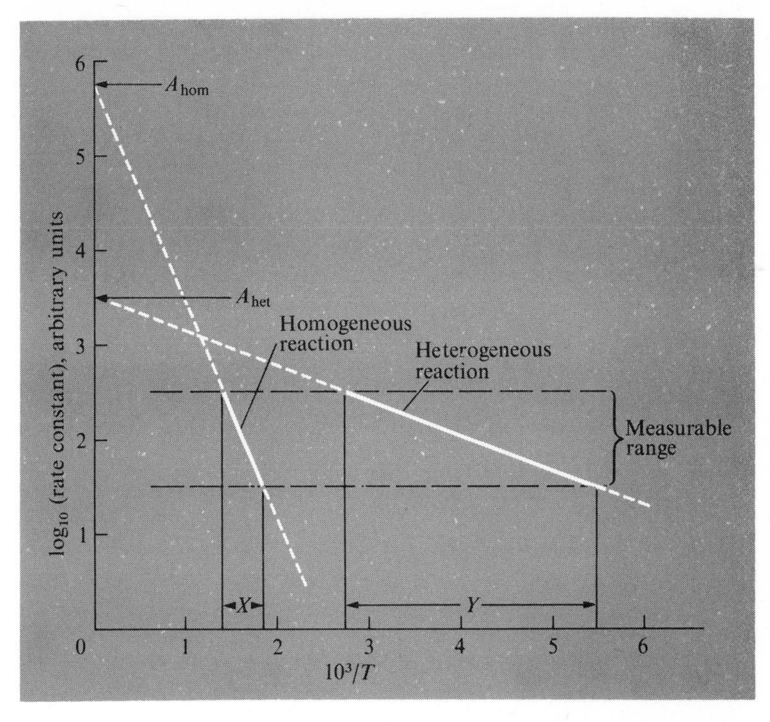

FIG. 1.3. Arrhenius diagram for uncatalysed and catalysed reactions: X and Y are the ranges of $1/T$ in which the respective reactions can be observed.

1.5. Implications of the definition of catalysis

There are other very important implications of our definition of catalysis, and these must now be examined. The definition implies that the position of equilibrium attained in the presence of a catalyst is the same as that ultimately arrived at when no catalyst is present. The laws of thermodynamics of course require this. They could not tolerate a situation in which two different positions of equilibrium, and hence two different free energy changes, could coexist. Thermodynamics, as we have seen above, concerns itself only with differences between initial and final states, and thus not with whether or not a catalyst is used to bring the change about. At any given temperature, therefore, the standard entropy and enthalpy changes must also be the same for both catalysed and uncatalysed reactions. This fact has been turned to good advantage in the measurement of enthalpy changes, that is to say, heats of reaction or heats of formation. The calorimetric measurement of a heat of reaction is much easier at temperatures close to ambient than at the much higher temperatures at which uncatalysed reactions usually proceed. Measurements made on a catalysed reaction proceeding at a conveniently low temperature yield equivalent information. An example of this is the determination of heats of hydrogenation. Table 1.2 shows that the heat of hydrogenation of ethylene determined calorimetrically with the aid of a catalyst is in excellent agreement with values derived from heats of combustion and from the temperature variation of the equilibrium constant of the uncatalysed system at high temperatures.

TABLE 1.2

Values for the heat of hydrogenation of ethylene (kJ *mol*$^{-1}$) *by various methods*

Method	$-\Delta H_h$
Calorimetric measurement at 355 K	136·3
Calculated from heats of combustion	137·1
Calculated from equilibrium constant at about 670 K	136·6

The foregoing remarks are of course valid only if the catalysed and uncatalysed reactions are truly the same, and yield identical products. It is indeed surprisingly rare to be able to make direct and quantitative comparisons between the two. Perhaps the easiest way of showing this is to consider a specific example. The homogeneous oxidation of propylene yields carbon dioxide and water as the only major products, but in the catalysed oxidation a great variety of intermediate oxidation products may be obtained in good yield: depending on the catalyst used, these may be acrolein, acrylic acid, acetone, acetaldehyde, acetic acid, and others. This point sometimes causes difficulties and it is worth spending a moment to consider how this should be.

It must first be emphasized, and this is another implication of our definition, that a catalyst can only increase the rate of a reaction which is already thermodynamically feasible: it cannot initiate a reaction which is thermodynamically impossible. Catalysed reactions are as much subject to the laws of thermodynamics as are uncatalysed reactions. If therefore the oxidation of propylene is capable of yielding intermediate oxidation products, then each reaction must be thermodynamically acceptable. They may however be isolated only if the contact time of the reactants with the catalyst is relatively short: they would certainly suffer complete oxidation to carbon dioxide and water were they to remain in contact with the catalyst indefinitely. They are in a sense metastable products. The situation may be represented as in Fig. 1.4, which is oversimplified, as there may of course be a succession of intermediate products formed sequentially, each being more stable than its precursor. There is therefore no fundamental reason why a catalysed reaction should give the same *initial* products as a non-catalysed reaction, although the *final* products will

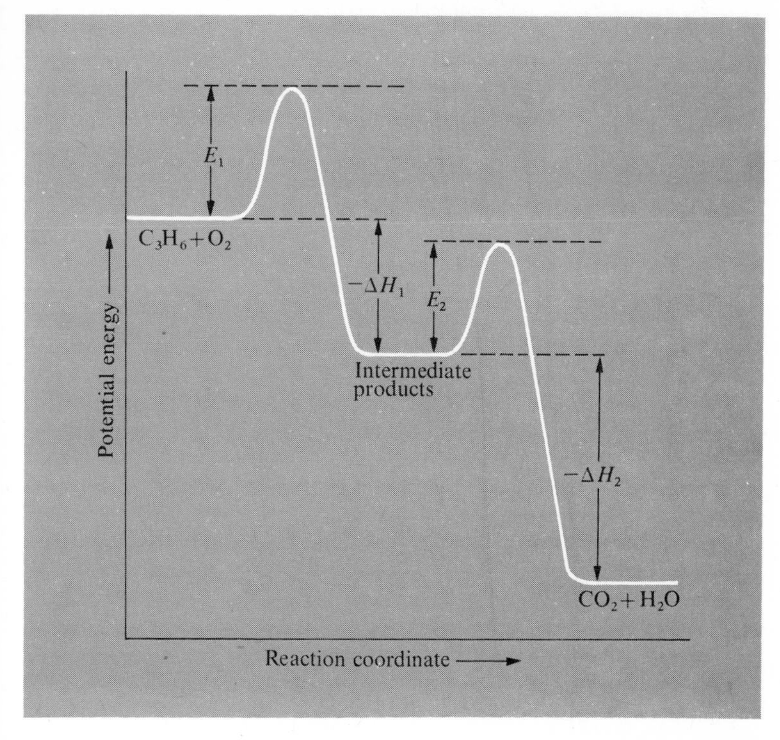

FIG. 1.4. Potential energy profile for a multistage exothermic reaction.

always be identical since for any reaction there can be only one set of stable products.

Since the presence of a catalyst does not alter the equilibrium constant, the quotient k_f/k_r must also remain unchanged. The catalyst does however increase k_f: it must therefore also increase k_r by the same factor. This statement is restricted to simple chemical transformations in which thermodynamically unstable intermediates do not occur. Use can however sometimes be made of this point: thus for example in searching for an improved ammonia synthesis catalyst, it is easier to study the decomposition of ammonia because at a suitably high temperature the equilibrium will favour the products even at atmospheric pressure.

A final word is in order concerning the significance of a catalyst's decreasing the activation energy of a reaction. This must mean that the catalysed reaction proceeds by a new and energetically more favourable pathway. How catalysts manage to achieve this will command our attention in later pages.

1.6. Classification of catalytic systems

The foregoing remarks apply indiscriminantly to all forms of catalysis: they are of universal validity where catalysis occurs. In fact it is possible to divide catalytic systems into two distinct categories. When the catalyst is of the same phase as the reactants and no phase boundary exists, we speak of *homogeneous catalysis*. This may take place either:

(i) in the gas phase, as for example when nitric oxide catalyses the oxidation of sulphur dioxide; or

(ii) in the liquid phase, as when acids and bases catalyse the mutarotation of glucose; or

(iii) in the solid phase, as when manganese dioxide catalyses the decomposition of potassium chlorate.

However, when a phase boundary does separate the catalyst from the reactants, we speak of *heterogeneous catalysis*. A number of phase combinations can then occur, as shown in Table 1.3. Other possible phase combinations rarely arise in catalysis.

TABLE 1.3

Phase combinations for heterogeneous catalysis

Catalyst	Reactant	Example
Liquid	Gas	Polymerization of alkenes catalysed by phosphoric acid.
Solid	Liquid	Decomposition of hydrogen peroxide catalysed by gold.
Solid	Gas	Ammonia synthesis catalysed by iron.
Solid	Liquid + Gas	Hydrogenation of nitrobenzene to aniline catalysed by palladium.

There is however one extremely important group of substances which cannot be accommodated within this classification. *Enzymes* are neither homogeneous nor heterogeneous catalysts; they are large, complex organic molecules, usually containing a protein, which form a lyophilic colloid: this is neither a homogeneous nor a heterogeneous system, but something in between. We must therefore regard enzymatic catalysis as something quite different from the other two forms of catalysis. However it is not within the scope of this book to discuss enzymatic catalysis in any detail.

QUESTIONS

1.1. Calculate the equilibrium concentrations of reactants and product for the system

$$A+B \rightleftharpoons C$$

where K has values of 0·5 and 5, and plot the results in the manner of Fig. 1.1.

2. The descriptive chemistry of heterogeneous catalysis

2.1. Classification of solid catalysts

WHEN the industrial scientist is called upon to devise a catalyst for a new application, he must have the basic information he needs in a readily accessible form. As the process of catalyst design becomes progressively more sophisticated with the increasing standards that are placed on a successful catalyst, so the chemical and physical complexity of the finished product becomes greater. If we may know in general terms the most important capabilities of various classes of substances, it becomes easier to select the ingredients for an effective catalyst.

Catalysis is essentially a chemical phenomenon. The ability of a substance to act as a catalyst in a specified system depends on its chemical nature. With heterogeneous catalysis we are concerned with the specific chemical properties of the surface of the chosen substance. These of course reflect the chemistry of the bulk solid, and some useful insight into the catalytic activities of surfaces is gained from knowledge of the bulk properties of the solid. Our first objective must be to try to understand in general terms how the type of reaction best catalysed by a solid depends on its chemical nature.

Table 2.1 presents a preliminary and somewhat superficial classification of solids into groups depending on their catalytic abilities. Fine structure within

TABLE 2.1

Classification of heterogeneous catalysts (less important functions in brackets)

Class	Functions	Examples
Metals	hydrogenation dehydrogenation hydrogenolysis (oxidation)	Fe, Ni, Pd, Pt, Ag
Semiconducting oxides and sulphides	oxidation dehydrogenation desulphurization (hydrogenation)	NiO, ZnO, MnO_2, Cr_2O_3, Bi_2O_3–MoO_3, WS_2
Insulator oxides	dehydration	Al_2O_3, SiO_2, MgO
Acids	polymerization isomerization cracking alkylation	H_3PO_4, H_2SO_4, SiO_2–Al_2O_3

each group will be considered later. This table may be in part interpreted using the qualitative concept of *compatibility* between catalyst, reactants, and products. For catalysis to occur, there must be a chemical interaction between catalyst and the reactant–product system, but this interaction must not change the chemical nature of the catalyst except at the surface. Thus by compatibility we mean the existence of a surface interaction which does not penetrate into the interior of the solid.

Table 2.1 shows that transition metals are especially good catalysts for reactions involving hydrogen and hydrocarbons. This is because these substances readily *adsorb* at the surfaces of metals, in a manner to be described in more detail below, and except in a few cases the reaction does not proceed below the surface. Base metals are useless as catalysts for oxidation because at the necessary temperature they are rapidly oxidized throughout their bulk. Only those 'noble' metals (such as palladium, platinum and silver) that are resistant to oxidation at the relevant temperature may be used as oxidation catalysts. Many oxides on the other hand are excellent oxidation catalysts because they interact with oxygen and other molecules, but with some important exceptions (e.g. copper chromite) they are not well suited for hydrogenation because of the likelihood of reduction to metal. Those oxides which may be used for hydrogenation or dehydrogenation are of course immune to reduction by hydrogen at the temperature at which they are active. Similarly metal sulphides catalyse reactions of molecules containing sulphur: if oxides are used, they rapidly become sulphided. Those oxides such as alumina, silica, and magnesia, which do not interact much with oxygen are poor oxidation catalysts, but they easily adsorb water and thus may be used to catalyse dehydration. The concept emerges that we must first understand the adsorption of molecules at solid surfaces before we can proceed to a deeper knowledge of catalysis.

2.2. Adsorption of molecules at solid surfaces

The adsorption of molecules at the surfaces of solids is a long-known and much studied phenomenon. The historical background is well documented and will not be covered in detail here. It has also long been recognized that the interaction we call adsorption comprises two entirely different processes, involving quite different forces.

Let us consider first for simplicity the formation of a surface by fracture of a crystal of a covalent solid, such as diamond or any metal. In this process covalent bonds between atoms are broken, and so each surface atom must possess one or more free valencies. The number and types of these valencies depends on the bonding between atoms in the bulk solid and the angle through the crystal which the fracture takes. An atom in the new surface is clearly in a rather unusual position in that it does not have its full complement of neighbours: its coordination number is smaller than for atoms within the bulk of the solid. Fig. 2.1, which illustrates diagrammatically the simplest possible situation, also shows that there is an imbalance of forces at the surface, and that a surface atom suffers a net force acting inwards. This gives rise to the phenomenon of surface energy, which is related to surface tension in liquids, but is much greater owing to the stronger forces involved. One form of adsorption occurs as a result of a molecule's interaction with these free valencies: this may be regarded as a chemical reaction because there is a rearrangement, sometimes drastic, of the electrons within the molecule. This type of adsorption is therefore termed chemical adsorption or *chemisorption*. Some further characteristics of this process will be noted later.

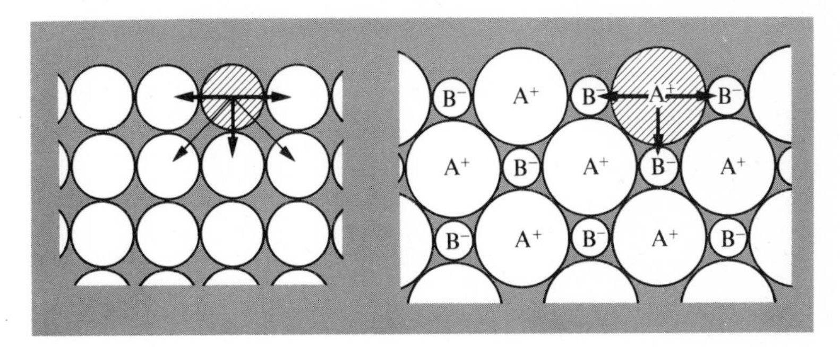

FIG. 2.1. Pictorial representation of the surface energy of a covalent solid.

FIG. 2.2. Pictorial representation of the surface energy of an ionic solid.

A precisely analogous situation occurs at the surface of ionic solids, as depicted in Fig. 2.2. Here there is also an imbalance of forces because each ion is surrounded by fewer than the proper number of ions of the opposite charge. Chemisorption on such surfaces attempts to rectify this situation, and in consequence it is often ionic or at least highly polar in character. Molecules whose electrons are not easily polarizable are only weakly adsorbed on ionic solids.

The other form of adsorption which is recognized to occur at the surfaces of solids is due to van der Waals forces such as exist between molecules themselves in the liquid state. These include electrostatic attraction in the case of molecules with permanent dipole moments, and induced dipolar attraction with readily polarizable molecules; dispersion forces caused by slight fluctuations in electron density are the only forces of attraction between non-polar atoms or molecules (see P. W. Atkins's *Quanta* (OCS 21) for further details). When these forces are exerted between an atom or molecule and a surface, there is a physical attraction without chemical alteration of the molecule: this is termed *physical adsorption* or less often physisorption. The strength of physical adsorption will clearly relate to the observable physical properties of the adsorbing species. Thus for example the adsorption of the inert gases and of nitrogen and hydrogen will be weak and detectable only at low temperatures, whereas, in the cases of say water and benzene, adsorption will be stronger and observable above 373 K. There is one other important point concerning physical adsorption: it does not depend much on the chemical nature of the solid. We can observe and utilize the low-temperature adsorption of nitrogen on fine metal powders and on porous silica with equal facility.

Chemisorption and physical adsorption are usually distinguishable from each other without any great difficulty. Table 2.2 summarizes the main criteria which can be applied for this purpose, and these will now be briefly discussed.

TABLE 2.2

Criteria for distinguishing between chemisorption and physical adsorption

Criterion	Chemisorption	Physical adsorption
Heat of adsorption, $-\Delta H_{ads}$	40–800 kJ mol^{-1}	8–20 kJ mol^{-1}
Activation energy, E	Usually small	Zero
Temperature of occurrence	Depends on E, but usually low	Depends on boiling point, but usually low
Number of layers adsorbed	Not more than 1	More than one possible

Adsorption is usually exothermic. This is so because, being a spontaneously occurring process, ΔG is negative and hence $(\Delta H - T\Delta S)$ must also be negative, or equally $T\Delta S$ must be larger than ΔH. Now ΔS will also be negative because adsorption produces a more ordered system with fewer degrees of freedom: hence ΔH must also be negative, and the process must therefore be exothermic. The molar heat of adsorption is the heat liberated when a mole of a substance is transferred from the gaseous to the adsorbed state. For chemisorption, values are in the range which is typical of chemical reactions and generally fall between 40 and 800 kJ mol^{-1}. For physical adsorption, values are usually in the range found for heats of liquefaction or vaporization and are typically less than 20 kJ mol^{-1}. Higher values are sometimes observed, especially when a highly polar molecule such as water adsorbs on an ionic solid: in such a case it may be difficult to define the kind of adsorption occurring.

The new bonds formed in chemisorption at a metal surface are always to some degree dipolar because of the electronegativity difference between the atoms forming them. This produces a slight increase or decrease in the number of conduction electrons in the solid, which may be sensed by measuring the change in electrical conduction consequent upon adsorption. This is particularly easily done in the case of adsorption on metal wires. Physical adsorption produces no such electrical effects.

Chemisorption leads at most to monolayer coverage of the available surface. With physical adsorption, several adsorbed layers may be formed, especially near the normal boiling point of the adsorbing substance (see section 4.5).

A further point of distinction between the two types of adsorption concerns their rates. The rate of physical adsorption is always fast, as is the condensation of a vapour on the surface of its own liquid, because there is no activation energy. The process of chemisorption, on the other hand, has an activation energy: the basic kinetic equation is

$$k_c = \sigma Z \exp(-E_c/RT)$$

where k_c is the rate constant for chemisorption, σ is the sticking probability, Z is the number of collisions of the adsorbing molecule per cm^2 of surface in unit time (this is proportional to gas pressure), and E_c is the activation energy for chemisorption. This equation is exactly analogous to that for a normal chemical reaction, based on the ideas of the collision theory (see p. 4). The value of σ is sometimes very low: on extremely clean plane metal surfaces at low pressure it can be as low as 10^{-6}, but of course Z is very large ($\sim 10^{23}$ $cm^{-2}\,s^{-1}$) and so the number of effective collisions per unit time is also quite high. Activation energies are generally low, so that the exponential Boltzmann factor approaches unity, and thus in the vast majority of cases surface coverage is complete in a fraction of a second.

In practice it is not easy to distinguish the two types of adsorption by their rates, since either can be immeasurably fast. A slow rate may signify an activated chemisorption, but equally it may well be caused by a slow diffusion of a physically adsorbing molecule into a porous adsorbent. Great care is needed when interpreting rates of adsorption.

For a molecule to react catalytically at a solid surface, it must first be chemisorbed. When two molecules so react, at least one and probably both must be chemisorbed. Chemisorption is an essential step in the preparation of a molecule for reaction: a chemisorbed molecule sometimes resembles the product into which it will be transformed more than it does the free molecule. It has also been suggested recently that chemisorption is equivalent to raising the molecule to its first excited state. Physical adsorption on the other hand has little relevance to catalysis.

2.3. Adsorbed states of molecules on metals

We must now give some thought to the mechanisms by which chemisorption occurs. It is at once apparent that many molecules are unable to react with the 'free valencies' of metal surfaces without undergoing a radical disruption of the bonds within them. In the case of hydrogen for example it is now very well established that to chemisorb it must dissociate into hydrogen atoms. This is represented as

$$H_2 + 2M \longrightarrow 2HM$$

where M signifies a surface metal atom. Saturated hydrocarbons fall in the same category. The only way in which methane can chemisorb is as

$$CH_4 + 2M \rightarrow CH_3M + HM.$$

Molecules which chemisorb in this way are said to be *dissociatively* chemisorbed. Molecules possessing π-electrons or lone-pair electrons can however chemisorb without dissociation. Thus for example mono-olefins are thought to chemisorb by a rehybridization of molecular orbitals in which the carbon atoms change from sp^2 to sp^3: in this way two free valencies are generated which can react with the free valencies of the metal. For ethylene we may therefore write

$$C_2H_4 + 2M \rightarrow \begin{array}{cc} H_2C & \!\!-\!\! CH_2 \\ | & | \\ M & M \end{array}$$

and for carbon monoxide

$$CO + 2M \rightarrow \begin{array}{ccc} & O & \\ & \| & \\ & C & \\ / & & \backslash \\ M & & M \end{array}$$

Such molecules are said to be *associatively* chemisorbed. Deeper discussion of the possible structures of chemisorbed molecules requires a description of the metal surface valencies in molecular-orbital language, but this cannot be attempted in this book. It must suffice to say that many metals seem to have empty orbitals at their surfaces and that these can accept electrons, so that for example the chemisorption of hydrogen sulphide, a potent catalyst poison, can be depicted as

$$H_2S + M \rightarrow \begin{array}{c} HSH \\ \downarrow \\ M \end{array}.$$

It is remarkably difficult to obtain direct and unequivocal information concerning the structure of chemisorbed molecules by the application of those techniques which have been so successfully applied to free molecules. The low concentration of the species and the presence of the metal are chiefly to blame. The method which has proved most useful is infrared spectroscopy: this has been widely applied to carbon monoxide, olefins, and many other molecules. The use of electrical conductivity change to deduce the polarity of the chemisorption bond has already been noted. For information on other techniques, the more advanced texts listed in the general bibliography must be consulted. One word of warning is however essential at this stage. Physical methods of examining the structure of chemisorbed layers describe only the species to be found there most abundantly: those which are intermediates in the catalytic processes may not be sensed at all if they are in the minority. The conclusions derived from methods such as infrared spectroscopy and low-energy electron diffraction must therefore be regarded circumspectly.

2.4. Potential-energy curves for adsorption

We are now ready to bring together much of the foregoing information through a treatment of adsorption in terms of potential-energy curves. The procedure, which we exemplify by considering the adsorption of hydrogen on nickel, is nevertheless a very general one. Having read this section, the reader might then try to draw his own diagrams for some of the other adsorption systems already mentioned.

The composite diagram is shown in Fig. 2.3. The horizontal line represents the zero of potential energy, and a molecule far from the surface is at this level. To get above this line, energy has to be supplied; on falling below it, energy is liberated. We are concerned to see how the hydrogen molecule's potential energy changes as it approaches the nickel surface, indicated by the vertical line at the left of the diagram. The process of physical adsorption is represented by the curve P. The heat of physical adsorption $-\Delta H_p$ is small, and there is no activation energy: the minimum is quite far from the surface, the distance from the plane through the nickel nuclei being approximately

$$\begin{aligned} & r_{Ni} + r_{Ni,\,vdW} + r_H + r_{H,\,vdW} \\ &= 0{\cdot}125 + 0{\cdot}08 + 0{\cdot}035 + 0{\cdot}08 \\ &= 0{\cdot}32\ \text{nm} \end{aligned}$$

where r_{vdW} represents the van der Waals radius of the indicated species.

The curve C is for chemisorption. It represents the process

$$2Ni + 2H \longrightarrow 2NiH$$

and for this reason it begins at a height D_{HH} (434 kJ) above the potential-energy zero. As the pair of atoms approach the surface, they are stabilized by the formation of the chemisorption bond, and the potential energy therefore falls: the depth of the minimum $-\Delta H_c$ reflects the heat of chemisorption, which is about 125 kJ mol^{-1} at low surface coverage, and its distance from the surface as defined is given by

$$r_{Ni} + r_H = 0{\cdot}125 + 0{\cdot}035 = 0{\cdot}16\ \text{nm}.$$

We now see that the activation energy for chemisorption E_c arises because the two curves intersect *above* the potential-energy zero: its magnitude will depend critically on the distance of the minima from the surface, and thus also on the atomic radii of the adsorbing atom and molecule. However chemi-

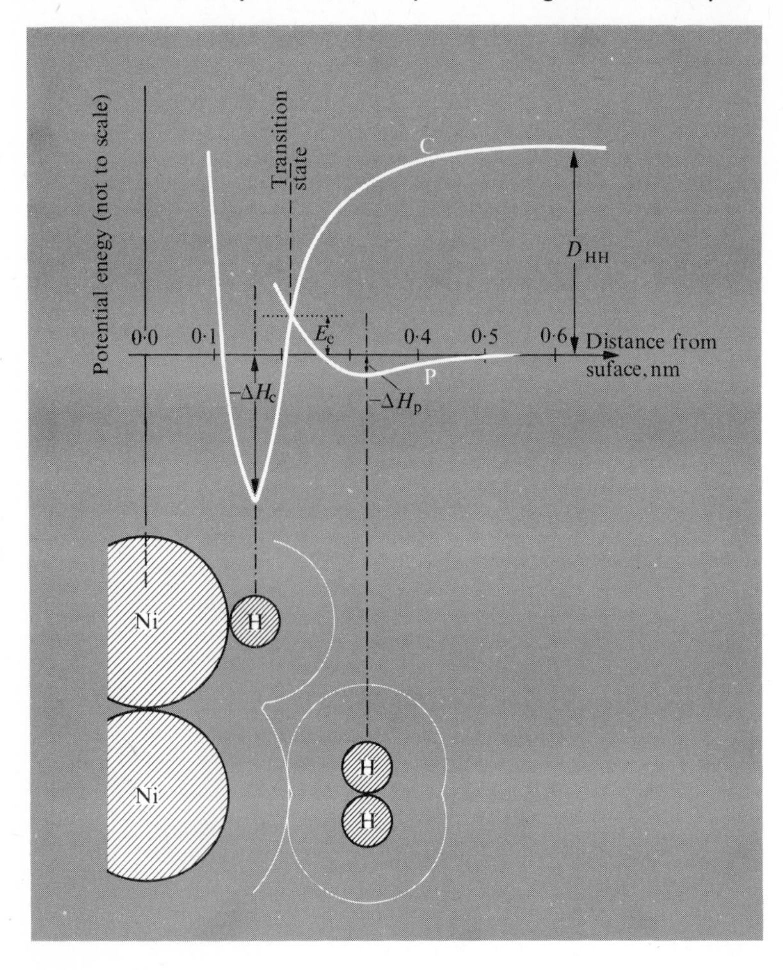

FIG. 2.3. Potential energy curves for the adsorption of hydrogen on nickel, and pictorial representation of the adsorbed states.

sorption can occur without providing the hydrogen molecule with the energy necessary to atomize it: this is just as well because the required energy is very large indeed, and a high temperature would be needed to produce hydrogen atoms in sufficient concentration for chemisorption to occur rapidly. The utility of physical adsorption is that *it permits the hydrogen molecule to get quite close to the surface without having to acquire much energy*: then at the intersection of the curves the transition from physical adsorption to chemisorption occurs. This is the *transition state* for chemisorption (see Fig. 2.4).

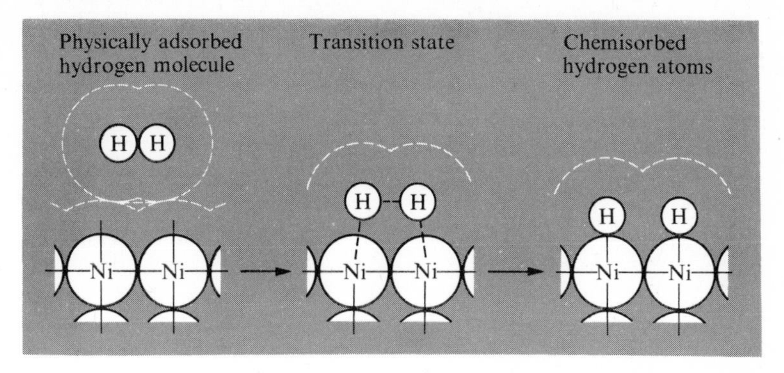

FIG. 2.4. Chemisorption of a hydrogen molecule on a nickel surface.

2.5. Descriptive chemistry of chemisorption on metals

For the detailed understanding of a catalytic reaction mechanism, knowledge of the structure and stabilities of adsorbed intermediates is of course most important. However for some purposes a simple qualitative indication of the occurrence or otherwise of chemisorption is sufficient. We must now consider the body of information on the chemisorption of various gases on metals: this constitutes the descriptive chemistry of chemisorption. With the aid of this information we shall be able to rationalize the catalytic properties of metals.

The chemisorption of a number of simple gases of practical interest has been investigated on many metals. It is found that, on the vast majority of metals, their strengths of adsorption fall in the following sequence:

$$O_2 > C_2H_2 > C_2H_4 > CO > H_2 > CO_2 > N_2.$$

Gold is an apparent exception to this generalization, because of its inability to chemisorb oxygen. Some metals whose surfaces are very reactive will chemisorb all these gases, including nitrogen, which is the most difficult of the molecules to activate. Others at the opposite extreme can chemisorb only oxygen, which is the easiest molecule to activate. Metals may therefore be categorized on the basis of the number of these gases they can chemisorb: this is done in Table 2.3. The qualitative nature of this table needs emphasizing. The criterion for chemisorption is merely whether it is detectable volumetrically at a pressure of about 10^2 Nm^{-2} (10^{-3} atm) and ambient temperature. In a few cases, the classification of a metal can depend on its physical state: thus technical copper catalysts can chemisorb hydrogen weakly, but pure metallic copper cannot. This does not however affect the value of this table as a rationalizing concept.

TABLE 2.3

A classification of metals according to their abilities in chemisorption

Group	Metals	Gases						
		O_2	C_2H_2	C_2H_2	CO	H_2	CO_2	N_2
A	Ti, Zr, Hf, V, Nb, Ta, Cr, Mo, W, Fe, Ru, Os	+	+	+	+	+	+	+
B_1	Ni, Co	+	+	+	+	+	+	−
B_2	Rh, Pd, Pt, Ir	+	+	+	+	+	−	−
B_3	Mn, Cu	+	+	+	+	±	−	−
C	Al, Au	+	+	+	+	−	−	−
D	Li, Na, K	+	+	−	−	−	−	−
E	Mg, Ag, Zn, Cd, In, Si, Ge, Sn, Pb, As, Sb, Bi	+	−	−	−	−	−	−

(+ means that strong chemisorption occurs; ± means that it is weak; − means unobservable.)

We must now briefly consider *why* metals can be grouped in this way. Study of Table 2.3 shows that the metals of Class A occur in Groups IV, V, VI, VII, and VIII_1 of the Periodic Classification†; those of Class B_1 are the base metals of Groups VIII_2 and VIII_3, while those of Class B_2 are the noble metals of these Groups. Class B_3 contains two anomalous metals of the first long series (adsorption and catalysis are full of anomalies), and all the most weakly chemisorbing metals (Classes C, D, and E) come either before or after the transition series proper. The propensity to show strong chemisorption is therefore firmly associated with transition metals.

The further understanding of this classification of metals really requires an effective way of describing the electronic structure of surface metal atoms, and this is an extraordinarily difficult problem. We shall however assume that they are similar to atoms within the bulk of the solid. Now the characteristic of the transition metals is that they have one or more unpaired d-electrons in the outermost electron shell, and the weakly chemisorbing non-transition metals have only s or p valency electrons. It has been suggested that unpaired d-electrons are necessary to bind the adsorbing molecule to the surface in a weakly-held precursor state, from which it then passes to the final strongly-bonded state. The availability of this intermediate state may serve to reduce the activation energy for adsorption to a conveniently low value, while it remains prohibitively high on those metals not having unpaired d-electrons. For some reactive molecules such as carbon monoxide and oxygen the

† Group VIII_1 comprises Fe, Ru, Os; Group VIII_2 Co, Rh, Ir; Group VIII_3 Ni, Pd, Pt.

formation of the precursor is not usually necessary, and they therefore chemisorb on almost all metals.

Table 2.3 if properly used has considerable predictive value. On the assumption (not always true) that a catalytic reaction between two molecules requires them both to be first chemisorbed, we may readily select those metals expected to have activity for say ammonia synthesis (Class A) or the hydrogen–oxygen reaction (Classes A, B_1, B_2, and perhaps B_3). This exercise is capable of extension, but the previous warning concerning the importance of kinetically significant adsorbed species in very low concentration must always be remembered.

2.6. Chemisorption and catalysis by metals—quantitative aspects

The foregoing treatment has been purely qualitative, and has not helped us to find out which metal is the *best* catalyst for a given reaction. To answer this question we need to propose a theorem relating the *strength* of chemisorption of the reactants on a metal with that metal's activity in catalysing reaction between them. The following generalization is advanced: that in a unimolecular reaction *the catalytic activity is inversely related to the strength of chemisorption of the reactant, providing that adsorption is sufficiently strong for the adsorbate to achieve high surface coverage.* If the reactant is very strongly adsorbed, it will clearly be unreactive, as the chemisorption bond will be too strong to be easily broken: this is the case with those many substances which act as *catalyst poisons.* If on the other hand the molecule is so weakly adsorbed that it covers only a very small fraction of the surface, then the catalytic rate will be at best minimal. This situation is shown diagrammatically in Fig. 2.5.

The position becomes a little more complicated when we come to consider the more realistic case of two reactants: in general these will not be equally strongly chemisorbed, and we may assume that the rate will be proportional to the product of the fractional surface coverage of each, viz.

$$r = k\theta_A\theta_B,$$

where θ_A is the coverage by A and θ_B is the coverage by B. Thus when the two reactants between them fully cover the surface, the rate will be maximum when θ_A equals θ_B, and will depend critically on variation of whichever is the smaller. At the same time however the strengths of adsorption of both reactants will be reflected in the magnitude of the rate coefficient, since this is concerned with the energy of the transition state formed from the two adsorbed molecules, quite irrespective of their concentrations. These two effects are separable, provided the kinetics of the reaction are known, but this is not always the case. More usually we are presented with a rate, without detailed knowledge of the values of reactant surface coverages.

Notwithstanding possible interpretative difficulties, we should now review the available results on the strengths of chemisorption of gases on metals, and

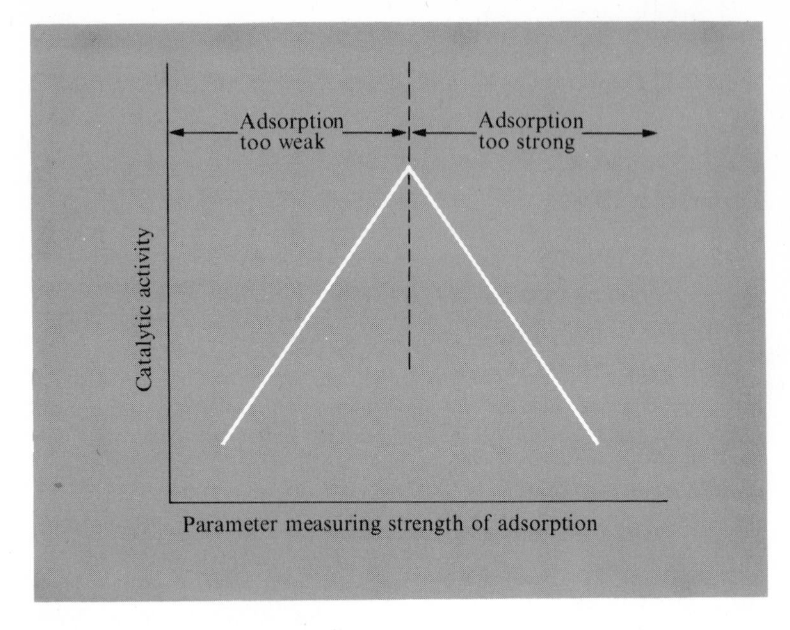

FIG. 2.5. Diagrammatic representation of a 'Volcano curve'.

the related catalytic activities. A suitable measure of chemisorption strength is the experimentally measured heat of adsorption, and extensive values are available. However, because heats of adsorption frequently decrease with increasing surface coverage, and not always in the same way, it is convenient to use heats extrapolated to zero surface coverage, the so-called initial heat of adsorption, and this is what we shall do here. (Quantitative aspects are discussed further in Chapter 3.)

Figure 2.6 shows the initial heats of hydrogen chemisorption as a function of periodic group number. Most of the datum points refer to evaporated metal films, but those for ruthenium, iridium, platinum, cobalt, and copper are for these metals supported on silica: the results for the last two metals have considerable margins of uncertainty. We see that in general the strength of adsorption decreases on moving from left to right through the periodic table, reaching a minimum in Group VIII_2: this is what we would have expected from Table 2.3. The behaviour of manganese is anomalous in the extreme (see also Table 2.3), but this merely reflects this element's unexpected physical and chemical properties, which are associated with the stability of the half-filled d-shell. On these grounds alone we should expect cobalt to be the most active metal for hydrogenation catalysis in the first long series, and ruthenium, rhodium, and iridium in the second and third long series to be comparably

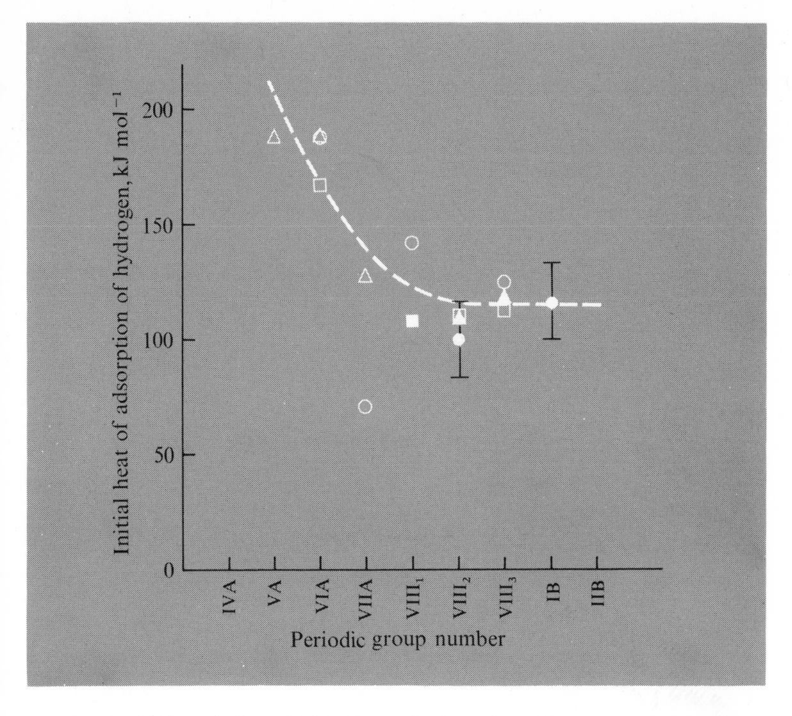

FIG. 2.6. Periodic variation of initial heat of adsorption of hydrogen. Open points, evaporated metal films: filled points, silica-supported metals: circles, first row transition metals: squares, second row transition metals: triangles, third row transition metals.

active. We should however take into consideration the adsorption of the other reactant before thinking about catalytic activities. Very many workers have used the hydrogenation of ethylene to ethane as a convenient measure of catalytic activity, but unfortunately ethylene adsorption has been looked at on only a few metals. However such values as are available for the initial heat of chemisorption of ethylene parallel closely those for hydrogen, although the values are some three times larger.

Rate coefficients for ethylene hydrogenation are plotted against periodic group number in Fig. 2.7. One or two uncertain values to be found in the literature (e.g. for iridium) are omitted. It seems strange that no one seems yet to have bothered to measure the activity of manganese in this reaction. The results are generally as expected; cobalt and copper are both less active than might have been expected, although our expectations may be in error for the reason noted above. The ensuing plot of activity against initial heat of hydrogen

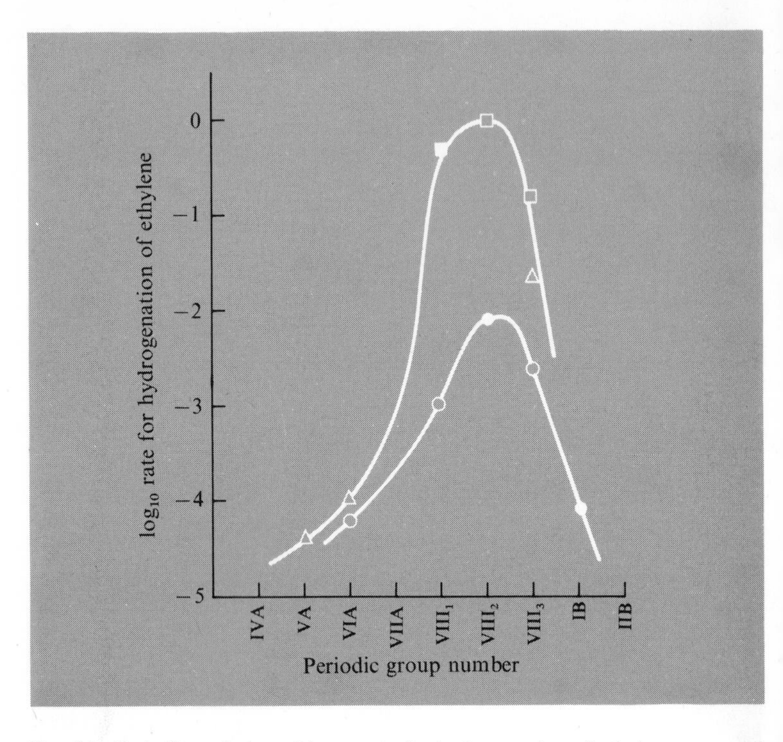

FIG. 2.7. Periodic variation of $\log_{10}$ rate for hydrogenation of ethylene expressed relative to rhodium. (Symbols as in Fig. 2.6.)

chemisorption is shown in Fig. 2.8. The uncertain positions of cobalt and copper are again highlighted: cobalt should almost certainly fall on the main curve, with a heat of adsorption of some 120 kJ mol^{-1}, but the rightful place of copper is more doubtful. We are in fact looking at the right-hand part of Fig. 2.5, and copper may be the only metal to reside on the left-hand part. Data now available are equivocal, and further work is required to resolve this matter.

The behaviour of metals as catalysts for ethylene hydrogenation is typical of their behaviour in most other hydrogenation reactions. The noble Group VIII metals are invariably the most active, although their relative activities differ importantly from one reaction to another: palladium for example is distinctly the most active metal for the hydrogenation of acetylene. There frequently are however over-riding economic considerations which lead to a less active metal such as nickel being used in industrial processes.

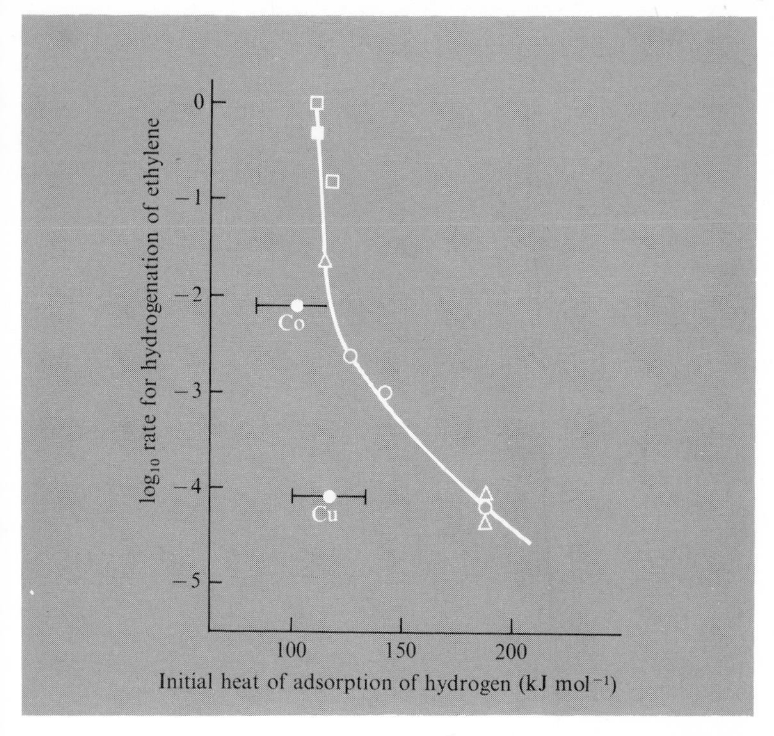

FIG. 2.8. Dependence of $\log_{10}$ rate for hydrogenation of ethylene (expressed relative to rhodium) on initial heat of adsorption of hydrogen. (Symbols as in Fig. 2.6.)

Another system replete with messages is the nitrogen–hydrogen combination: their reaction to form ammonia is one of the most important catalytic processes. The heat of adsorption of nitrogen has been measured on only a few metals (see Fig. 2.9) and ability to chemisorb nitrogen as atoms virtually disappears between Groups VIII_1 and VIII_2. Activity for ammonia synthesis is maximal in Group VIII_1 (iron is of course the catalyst used industrially) and decreases rapidly to either side (see also Fig. 2.9), either because of increasing strength of reactant adsorption or because of rapidly diminishing concentration of adsorbed nitrogen atoms. The metals of Groups VIII_2 and VIII_3 are presumably unable to chemisorb the trivalent nitrogen atoms because their surface atoms do not possess a sufficient number of unpaired electrons to hold them.

The problem of understanding the basis of the periodic variation in strengths of adsorption and in catalytic activity shown by metals is a highly complex one

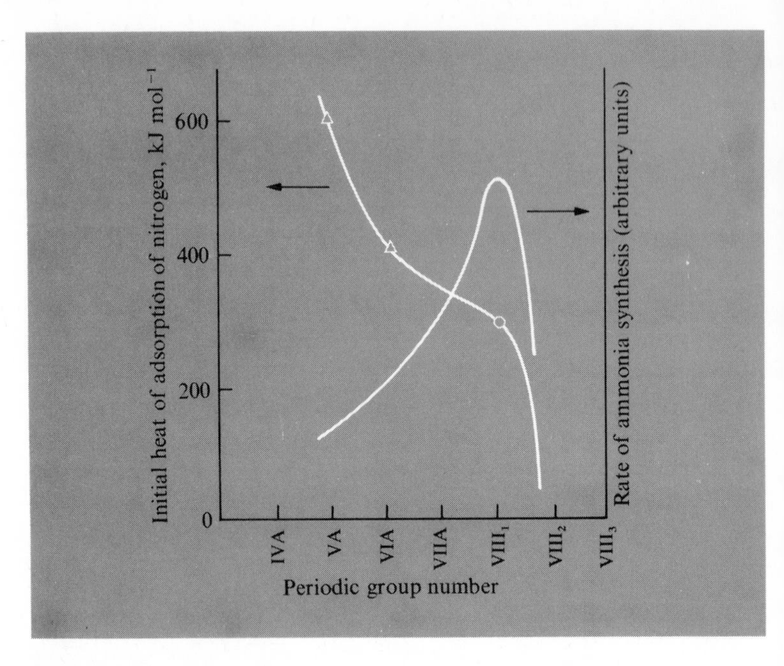

FIG. 2.9. Periodic variation of rate of ammonia synthesis (expressed qualitatively) and of initial heat of nitrogen adsorption (see also Fig. 2.8.).

which cannot now be treated in any depth. However examination of results on oxygen chemisorption throws a little further light. The periodic variation of initial heats of oxygen chemisorption is seen in Fig. 2.10, and more significantly these are plotted against the heat of formation of the most stable oxide in the next figure. The correlation is really quite a good one for this branch of science, and suggests that adsorbed oxygen atoms on the surface of a metal are broadly similar to oxide ions within the bulk of the corresponding oxide. Unfortunately there are very few bulk hydrides or nitrides whose heats of formation can be compared with the heats of adsorption of hydrogen and nitrogen, although perhaps fortuitously the point for the tantalum–nitrogen system lies almost on the line (see Fig. 2.11).

2.7. Adsorption and catalysis on semiconducting oxides

Semiconductors are substances whose electrical conductivity lies between 10 and 10^{-5} $(\Omega\,\mathrm{cm})^{-1}$ and increases with temperature. Some pure elements such as silicon and germanium, and compounds such as indium arsenide, show semiconductivity, but they are of no interest in catalysis and will not be considered further. Oxides are either semiconductors, or insulators having

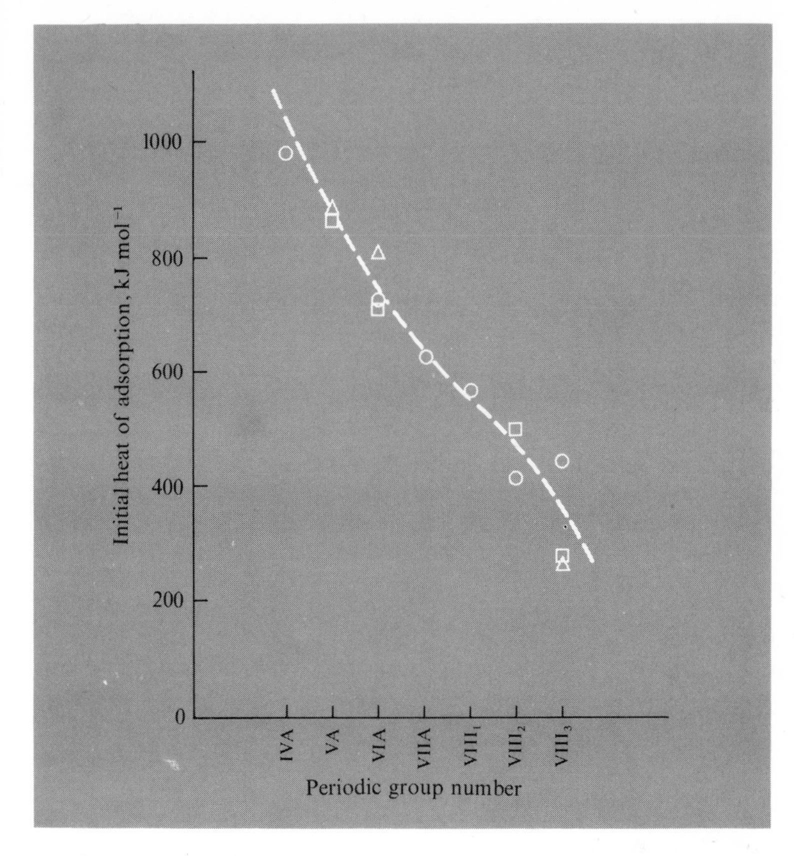

FIG. 2.10. Periodic variation of initial heat of adsorption of oxygen on evaporated metal films. (Symbols as in Fig. 2.6.)

electrical conductivities below 10^{-10} $(\Omega\,\text{cm})^{-1}$. As a rough guide, oxides of metals or metalloids in the short series (e.g. boron, silicon, magnesium, aluminium), and alkaline earth oxides, are always stoichiometric, have very high melting points, and are insulators. Oxides and sulphides of the transition and post-transition metals can become non-stoichiometric and when in this state are semiconductors. These oxides are in turn subdivided on the basis of whether they gain or lose oxygen when heated in air (see Table 2.4). We now attempt an explanation in chemical language of how non-stoichiometry in terms of oxygen excess or deficiency leads to p- or n-type semiconducting respectively. Physicists interpret the same phenomena in a quite different and perhaps more realistic way, using the Band Theory of the solid state—see

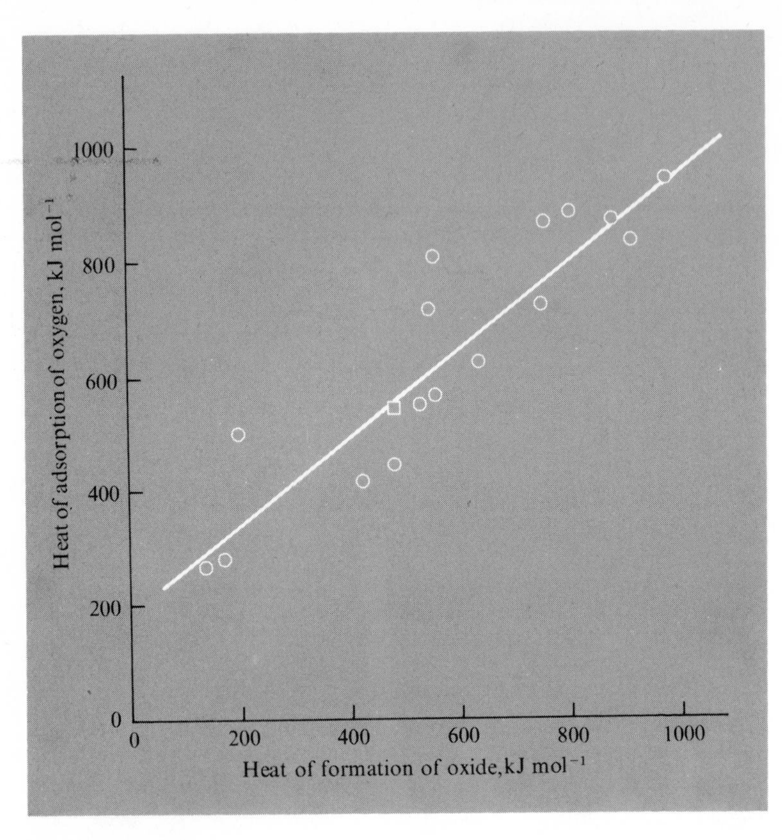

FIG. 2.11. Dependence of heat of adsorption of oxygen (see Fig. 2.10) on heat of formation of most stable oxide at 25°C. Square point is for the nitrogen–tantalum system.

TABLE 2.4

Classification of semiconducting metal oxides

Effect of heating in air	Classification	Examples
Oxygen lost	Negative (n-type)	ZnO, Fe_2O_3, TiO_2, V_2O_5, CrO_3, CuO
Oxygen gained	Positive (p-type)	NiO, CoO, Cu_2O, SnO, PbO, Cr_2O_3

Atkins: *Quanta* (OCS 21)—but the chemical approach is sufficient for our purposes.

Let us consider first the case of nickel(II) oxide which becomes non-stoichiometric in the sense of oxygen excess on heating in oxygen or air. Making additional oxide ions by the process

$$O_2 + 4e^- \rightarrow 2O^{2-}$$

requires electrons which can only come from Ni^{2+} ions, i.e.

$$Ni^{2+} \rightarrow Ni^{3+} + e^-.$$

The net process for oxygen acquisition is therefore

$$4Ni^{2+} + O_2 \rightarrow 4Ni^{3+} + 2O^{2-}.$$

The Ni^{3+} ions thus formed, two for each new oxide ion, are centres of excess positive charge or *positive holes*. Electron migration under the influence of a potential difference occurs by hopping from Ni^{2+} to Ni^{3+}: thus electron movement in one direction is accompanied by positive-hole migration in the opposite direction. Conductivity is proportional to positive-hole concentration and the mechanism is by positive-hole conduction: for this reason we speak of p-type semiconductivity.

Zinc oxide on the other hand becomes oxygen-deficient on heating in air. Loss of oxygen according to the process

$$2O^{2-} \rightarrow O_2 + 4e^-$$

generates electrons which reduce Zn^{2+} ion to zero-valent zinc atoms, the net process being

$$2Zn^{2+} + 2O^{2-} \rightarrow O_2 + 2Zn^{0}.$$

Electrical conduction is by migration of the electrons from the zinc atoms to Zn^{2+} ions, and conductivity is proportional to the concentration of zinc atoms: we therefore speak here of n-type (for negative) semiconductivity.

A useful generalization concerning the requirement for p-type semiconductivity is that the cation shall have an accessible higher oxidation state: thus cobalt(II) oxide and copper(I) oxide are also in this group. For n-type semiconductivity an accessible *lower* oxidation state (which may include the zero-valent state) is needed: thus cadmium oxide and iron(III) oxide fall in this group. It now becomes clear why the insulator oxides are to be found either before or at the very beginning of the transition series, for these have neither a higher nor a lower oxidation state which can be attained without massive energy input. The electrical conductivity of metal oxides is seen to be an inevitable consequence of their fundamental chemistry.

The chemisorption of gases is more complex on oxides than on metals. To start with, the adsorbed molecule may be attached either to a cation or to an oxide ion. Further, it has long been known that the adsorption of reducing gases such as hydrogen and carbon monoxide is substantially irreversible in the sense that, when desorption is attempted, only water and carbon dioxide respectively can be obtained. The process has clearly led to reduction of the surface, and it may be guessed that the adsorbed species resided or were associated at least in part with oxide ions. Oxygen adsorption is found to occur far more extensively on p-type oxides than on n-type oxides: this is because electrons may be withdrawn from Ni^{2+} ions (for example) to facilitate the formation of species such as O^-, whereas no such mechanism is available in the case of Zn^{2+} ions. On n-type oxides, oxygen adsorption really only occurs on pre-reduced surfaces, replacing oxide ions removed in a reducing pretreatment.

The catalytic properties of the two classes of oxides are substantially different, and the difference stems from their various abilities to adsorb oxygen. Let us consider first the decomposition of dinitrogen oxide according to the equation

$$2N_2O \rightarrow 2N_2 + O_2.$$

This reaction has been studied on many oxides, and it has been noted that p-type oxides are active between 470 and 570 K, whereas n-type oxides are only effective at 820 to 1020 K. Insulator oxides have intermediate activities. These observations may be rationalized as follows. The initial step in dinitrogen oxide decomposition is N–O bond fission brought about by the molecule's receiving an electron into an antibonding orbital. On nickel(II) oxide we could write

$$N_2O + Ni^{2+} \rightarrow N_2 + O^- \cdots Ni^{3+}.$$

This step is formally similar to oxygen chemisorption, and the high activity of p-type oxides in this reaction parallels their facile adsorption of oxygen. The adsorbed oxide ions are removed either by

$$2(O^- \cdots Ni^{3+}) \rightarrow O_2 + 2Ni^{2+}$$

or

$$N_2O + O^- \cdots Ni^{3+} \rightarrow O_2 + N_2 + Ni^{2+}.$$

The insulator and n-type oxides are much less active because of their inability to chemisorb oxygen.

The mechanisms whereby carbon monoxide is oxidized according to the equation

$$2CO + O_2 \rightarrow 2CO_2$$

also differ in the two groups of oxides. On the p-type oxides, adsorbed O^- reacts with adsorbed carbon monoxide as

$$(O^- \cdots Ni^{3+}) + CO \rightarrow CO_2 + Ni^{2+}.$$

On n-type oxides, carbon monoxide reacts with two oxide ions in the surface in the reductive process:

$$CO + 2O^{2-} \rightarrow CO_3^{2-} + 2e,$$

the latter probably reducing a Zn^{2+} ion to Zn^{o}. One of the two oxide ions is regenerated when the carbonate ion decomposes:

$$CO_3^{2-} \rightarrow CO_2 + O^{2-},$$

and the other is formed in the reoxidation of the surface:

$$\tfrac{1}{2}O_2 + Zn^{o} \rightarrow O^{2-} + Zn^{2+}.$$

The difference between these mechanisms, one involving adsorbed O^- and the other lattice O^{2-}, leads to profoundly different behaviours in more complex oxidations, as we shall see presently.

It is more difficult to find useful correlations between the catalytic activities of oxides and their chemical or physical properties than was so in the case of metals. Many of the transition metals exhibit variable valency and in such cases several oxides are possible, each with its distinctive structure, stability, and catalytic ability. Some of the cations at the surface may, as we have just seen, be in a higher oxidation state than those in the bulk: moreover of course there may be non-stoichiometry throughout the bulk. Semiconductivity type is at best only a rough guide to catalytic activity, since we are usually concerned just with the oxide surface, and at temperatures and in atmospheres quite different from those usually used in measurement of semiconductivity. p-Type oxides are however generally better catalysts for oxidation reactions than n-type oxides, because adsorbed oxygen species are more reactive than lattice oxide ions.

It is however possible to proceed a little further. If the breaking of the bond joining the relevant oxygen species to the surface is the rate-determining step in oxidation reactions, we should find an inverse correlation between catalytic rate and the oxygen-catalyst bond strength. This quantity has been obtained either from the temperature dependence of the oxygen dissociation pressure or from the heat of formation of the oxide per oxygen atom. Reasonably encouraging correlations have been found for the oxidation of hydrogen and of many hydrocarbons, both saturated and unsaturated, but they are still only approximate and based on too few examples.

2.8. The selective oxidation of hydrocarbons

The pure oxides which we have thus far considered are of little practical interest in hydrocarbon oxidation since they only catalyse 'deep' oxidation to carbon dioxide and water: this is especially true of p-type oxides. They are of course of interest in the control of atmospheric pollution by catalytic means. Of much greater interest and importance is the selective oxidation of various hydrocarbons which in the last two decades have become abundantly available at low price. Partially oxidized products are worth much more than the hydrocarbons they are derived from, and great effort has been given to the discovery and development of suitable catalysts for these selective oxidations. Some more general aspects of the petrochemicals industry will be described later (Chapter 8): here we concentrate on the principal factors of catalyst composition that determine the kind of products produced.

Catalytic reactions of hydrocarbons with oxygen can be classified as shown in Table 2.5.

TABLE 2.5

Classification of catalytic reactions of hydrocarbons with oxygen

Reaction without carbon–carbon bond fission

Without oxygen incorporation

Oxidative dehydrogenation (e.g. butene to butadiene)
Oxidative dehydrocyclization (e.g. hexane to cyclohexane)
Oxidative dimerization (e.g. propylene to hexadiene and benzene)

With oxygen incorporation

Formation of aldehydes and ketones (e.g. propylene to acrolein and acetone)
Formation of alcohols (e.g. propylene to allyl alcohol)
Formation of unsaturated acids (e.g. propylene to acrylic acid)
Formation of esters (e.g. ethylene and acetic acid to vinyl acetate)

Reaction with carbon–carbon bond fission

Formation of saturated aldehydes and acids (e.g. propylene to acetaldehyde and acetic acid)
Reaction of alkyl aromatics without ring fission (e.g. toluene to benzene)
Reaction of aromatics with ring fission (e.g. benzene to maleic anhydride, naphthalene to phthalic anhydride)

Deep oxidation

Formation of CO and CO_2

It is now necessary to attempt to summarize a very large body of work performed, mainly in industrial laboratories, over the past decade. Although many oxides are capable of showing some degree of selectivity in forming partially oxidized products, few if any show a high enough selectivity to warrant their use in an industrial process. A major breakthrough occurred in the late 1950s when it was discovered in the Sohio laboratories that a *compound oxide*, namely bismuth molybdate, showed an acceptably high selectivity in the oxidation of propylene to acrolein:

$$C_3H_6 + O_2 \rightarrow CH_2{=}CH{-}CHO + H_2O.$$

This was soon followed by the even more important finding, that propylene could be oxidized in the presence of ammonia in a one-step *ammoxidation* to give acrylonitrile:

$$C_3H_6 + NH_3 + \tfrac{3}{2}O_2 \rightarrow CH_2{=}CH{-}C{\equiv}N + 3H_2O.$$

Acrylonitrile is of course a monomer of great importance in the synthetic fibres industry. These discoveries initiated an immense programme of industrial research designed to uncover similarly valuable processes: many of the results are therefore to be found only in the patent literature.

The possible intermediate oxidation products from a simple olefin are several: thus for example propylene can yield allyl alcohol, acrolein, acrylic acid, and acetone, depending on whether reaction takes place at the terminal or central carbon atom, as well as acetaldehyde, acetic acid, and formaldehyde through carbon–carbon bond fission. 1,5-Hexadiene and benzene are formed by dimerization. Ethylene can give acetaldehyde and acetic acid, and hence a one-step synthesis of vinyl acetate, another useful monomer, is conceivable.

The mechanism of oxidation of propylene to acrolein over 'bismuth molybdate' is quite well understood. The propylene molecule first splits into a hydrogen atom and an allyl radical,

$$CH_2{=}CH{-}CH_3 \rightarrow CH_2{\cdots}CH{\cdots}CH_2 + H\cdot.$$

The allyl radical then reacts with a lattice oxide ion, another of which reacts with the two hydrogen atoms to form water

$$H\cdot + CH_2{\cdots}CH{\cdots}CH_2 + 2O^{2-} \rightarrow CH_2{=}CH{-}CHO + H_2O + 4e^-.$$

The electrons lower the oxidation state of the cations: the surface is then rapidly reoxidized to its former state by molecular oxygen. Hexadiene and benzene arise through the pairing of two allyl radicals.

The 'bismuth molybdate' system is somewhat complicated, there being at least four distinct compounds possible, depending on the Bi:Mo ratio. Not all are equally selective, the best being that having a Bi:Mo ratio of 2:1. It is now clearly recognized that easily available lattice oxide ions are required, and selectivity parallels other quantities (e.g. activation energy for hydrogen reduction) which reflect their availability. Deep oxidation to oxides of carbon is thought to involve *adsorbed* oxygen. In support of this, p-type oxides such as nickel oxide and manganese dioxide show very low selectivities while n-type oxides (MoO_3, Sb_2O_5, V_2O_5, ZnO etc.) show higher selectivities. Semiconductivity is not however a useful guide to understanding the behaviour of compound oxide systems; their thermochemistry and acidity seem to be far more important characteristics.

A useful generalization concerning the classification presented in Table 2.5 is that nearly all catalysts known to be effective for Class 1 reactions contain either molybdenum or antimony oxides, with one or more other components of great variety; those effective for Class 2 reactions usually contain vanadium pentoxide, with one or more additives. This observation awaits explanation.

For the sake of completeness we must note that it is not merely compound oxides that are good selective oxidation catalysts: two adjacent metals also show outstanding properties. Silver has long been known as an excellent catalyst for the oxidation of ethylene to ethylene oxide, while palladium is capable of oxidizing it to acetaldehyde (and thence in the presence of acetic acid to vinyl acetate) and also to acrylic acid. Other metals perform more poorly. It has recently been claimed that a new allotrope of silver, referred to as 'layer silver', oxidizes propylene to propylene oxide with fair selectivity: this is a reaction which had hitherto resisted catalysis.

2.9. Catalysis by aluminosilicates

A further interesting example of how two different oxides can cooperate to produce new phenomena is provided by the alumina–silica system. Neither oxide when pure has strong acidic character; when however a compound is made in which about 10 per cent alumina is dispersed throughout a silica matrix, marked acidity is observed. This arises in the following way. Silica may be thought of as being built up from SiO_4^{4-} tetrahedra, each oxygen being shared by two tetrahedra. Substitution of aluminium for silicon gives the AlO_4^{5-} ion, so for each aluminium atom introduced there is one excess negative charge which needs to be balanced: if this is done by a proton, the material is strongly acidic (see Fig. 2.12). Equivalently we may say that if the aluminium remains three-coordinate, a neighbouring SiO_4^{4-} will carry an excess negative charge which may be balanced by a proton in the form of an acidic hydroxyl group (see also Fig. 2.12). These acidic species can then catalyse hydrocarbon reactions that proceed by a carbonium-ion mechanism (see Chapter 7). Many other compounds containing atoms having different valencies show acidity for the same reason.

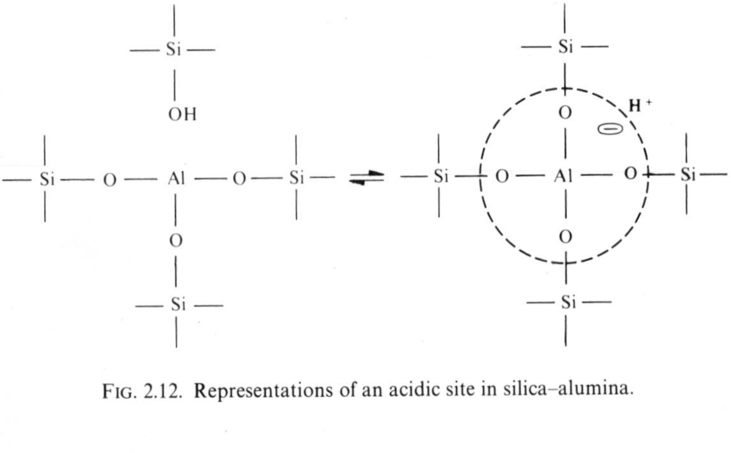

FIG. 2.12. Representations of an acidic site in silica–alumina.

Such aluminosilicates are amorphous, and much interest has been shown in recent years in crystalline aluminosilicates having the general formula $M_v(AlO_2)_x(SiO_2)_y{\cdot}zH_2O$: these are known as *zeolites*. When M is a monopositive cation (e.g. sodium or ammonium), v equals x; for divalent cations, v is $x/2$, and so on. There are a great many naturally occurring minerals of this kind, but the *mordenite* and *faujasite* classes are of greatest interest as catalysts. We shall confine our attention here to the latter, some members of which can now also be synthesized.

The basic unit of a faujasite is the regular cubo-octahedron or sodalite unit (see Fig. 2.13) consisting of 24 tetrahedra of either SiO_4^{4-} or AlO_4^{5-}. Two general types of structure can then arise, depending on the way in which the sodalite units are joined together. When they are joined through their square faces, we obtain the *A zeolite* in which y/x is unity (see Fig. 2.14); when through their hexagonal faces, the *X zeolite* (y/x, 1·25) or the *Y zeolite* (y/x, 1·5 to 3) (see Fig. 2.15). A feature of these last two structures is that they have quite

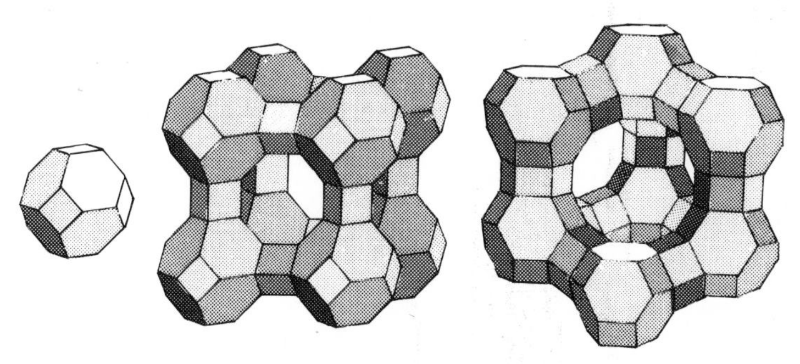

FIG. 2.13. The cubo-octahedral unit of the faujasite zeolites.

FIG. 2.14. Structure of A zeolite.

FIG. 2.15. X and Y zeolite structure.

large cages joined by smaller openings: they therefore have a very high internal surface area in the form of pores of fixed geometry. The molecules of water are readily removed by heating, whereupon they become excellent desiccating agents, but more importantly the size of the opening between the cages determines the size of adsorbing molecule which can gain access. The ability of a structure to adsorb some molecules and to reject others is the basis of *molecular sieve action.*

The effective pore diameter is determined by the kind of cation which balances the negative charge on the structure. Cations occupy one of three distinct positions, and their size thus controls the size of the openings. With A zeolite, the effective pore diameter is 0·3 nm when the cation is potassium,

0·4 nm when it is sodium, and 0·5 nm when it is calcium. Zeolites are usually first made in the sodium form: the sodium ions can then quite simply be exchanged for other cations. An ammonium zeolite on heating loses ammonia and leaves behind a proton: such a zeolite is said to be *decationated* but complete decationation does not appear possible.

The useful catalytic properties of zeolites hinge on three factors: (i) the regular crystalline structure and uniform pore size which allows only molecules below a certain size to react; (ii) the presence of strongly acidic hydroxyl groups which can initiate carbonium-ion reactions; and (iii) the presence of very large electrostatic fields in the neighbourhood of the cations which can thus induce reactivity in reactant molecules. Catalytic activity therefore depends heavily on the nature of cation, which also seems able to affect the acidity of the hydroxyl groups.

Reactions catalysed by zeolites are predominantly those proceeding by carbonium-ion mechanisms, such as skeletal rearrangements and cracking of hydrocarbons. Zeolites containing rare-earth ions are employed on an industrial scale in the petroleum industry for hydrocarbon cracking. Many other reactions can occur, including oxidation, dehydration, dehydrogenation, and (when a suitable cation such as Pt^{2+} or Ni^{2+} is present) hydrogenation. We shall hear much more about these complex but useful catalysts in the years to come.

QUESTIONS (*ADVANCED)

2.1. Sketch potential-energy diagrams for the chemisorption of nitrogen, ethylene, carbon monoxide, methane, and water on a metal. Show by means of diagrams (see Fig. 2.3) what forms you are assuming for the adsorbed states.

2.2. What oxides might be substituted for alumina to give acidic centres when dispersed in silica? What other combinations of oxides might exhibit acidity?

2.3. Referring to Table 2.3, into which group do you think rhenium might fall? Why?

2.4. From Fig. 2.6, estimate the initial heat of chemisorption of hydrogen on osmium.

2.5. With reference to Figs 2.6–2.8, have a guess at the activity of manganese for ethylene hydrogenation.

2.6. Which types of semiconductivity are likely to be shown by the following oxides: VO_2, Cu_2O, WO_3, MnO_2, Nb_2O_5?

2.7. According to Table 2.3, which elements should be unable to catalyse the oxidation of acetylene?

2.8. Propylene labelled with ^{13}C in the methyl group reacts with oxygen over bismuth molybdate to give acrolein: where in the acrolein molecule do you expect the labelled atom to appear?

2.9. Draw possible structures for the adsorbed states of oxygen, nitrogen, acetylene, ethane, benzene, phenol, and nitric oxide on a metal surface.

*2.10. In section 2.5, reference is made to 'a weakly held precursor state'. (i) Suggest possible forms for this state in the adsorption of hydrogen and of ethylene. (ii) How should the potential-energy curves (see Fig. 2.3 and your answer to question 1) be modified to allow for the existence of this additional state?

*2.11. How will the potential-energy curves describing adsorption on (a) a simple oxide and (b) a compound oxide differ from those applying to metals? Illustrate your answer by reference to hydrogen, oxygen, and propylene.

2.12. Make a model of a metal surface using plastic spheres, and study the adsorbed state of hydrogen and other molecules using 'atoms' made to scale from Plasticine or other convenient material.

3. Quantitative aspects of adsorption and catalysis

3.1 Adsorption isotherms

THUS far, most of our discussion of the phenomena of adsorption and catalysis has been qualitative, except insofar as it has concerned calorimetrically determined heats of adsorption or reaction rates. It is now desirable to adopt a more quantitative approach. For example, when a new catalytic process is being developed for industrial use it is most important to know the kinetics of the reaction so that the reactor may be optimally designed and the process made as efficient as possible. Since the catalytic process involves interaction between reactants adsorbed on the surface, we must begin by examining how their concentrations are related to gas-phase pressure.

When a quantity of gas is admitted to an outgassed solid in a vacuum, part of the gas is adsorbed on the surface and part remains unadsorbed. We are not concerned here with the *rate* of adsorption, which is usually very fast, and almost always faster than the ensuing chemical reaction. (An important exception to this generalization is ammonia synthesis, where under some conditions the adsorption of nitrogen is believed to be rate-limiting (see Chapter 10).) Our concern is with the situation after adsorption is complete, and equilibrium attained. The relation at constant temperature between the quantity of gas adsorbed and the pressure with which it is in equilibrium is termed an *adsorption isotherm*. Provided the solid is non-porous and the temperature is well above the boiling point of the gas, then the isotherm is usually of the form shown in Fig. 3.1. The adsorption may be followed either gravimetrically or volumetrically; if x is the amount adsorbed at a given pressure and x_{max} is the maximum amount which the surface can take up, then

$$x/x_{max} = \theta$$

where θ is the fractional surface coverage. We now develop a relation between θ and equilibrium pressure P.

Let us imagine that each square metre of surface consists of n unit *sites*, each of which can adsorb one, but only one, molecule of the gas. All sites are supposed to be energetically equivalent. The equilibrium which is finally set up is naturally a dynamic one, and the isotherm is derived by formulating the rates of adsorption and desorption, and then equating them.

The equilibrium is represented as

$$A + M \underset{k_d}{\overset{k_a}{\rightleftharpoons}} A\text{–}M$$

where A is the adsorbing molecule, M a surface site, and A–M the chemisorption complex: k_a and k_d are the rate coefficients for adsorption and desorption respectively. From the *law of mass action* we have

$$\text{rate of adsorption} = k_a[\text{A}][\text{M}]$$

and

$$\text{rate of desorption} = k_d[\text{A–M}].$$

Now [A] can be replaced by P, the pressure of A at equilibrium, and [M], which is the concentration of vacant sites, by $n(1-\theta)$: similarly [A–M] is given by $n\theta$. Therefore on equating the rates of adsorption and desorption, and making these substitutions, we obtain

$$k_a Pn(1-\theta) = k_d n\theta;$$

whence by rearrangement

$$\theta = \frac{k_a P}{k_d + k_a P} = \frac{bP}{1+bP}$$

where b is k_a/k_d and is called the *adsorption coefficient* of A on the solid being used. We note incidentally that the value of n is irrelevant. The adsorption coefficient is clearly the equilibrium constant, and its magnitude reflects the strength of adsorption of A: thus if b is large, the equilibrium lies well to the right, and A is strongly adsorbed. This relation between fractional surface coverage and pressure was first developed by Irving Langmuir and is usually known as the *Langmuir adsorption isotherm.*

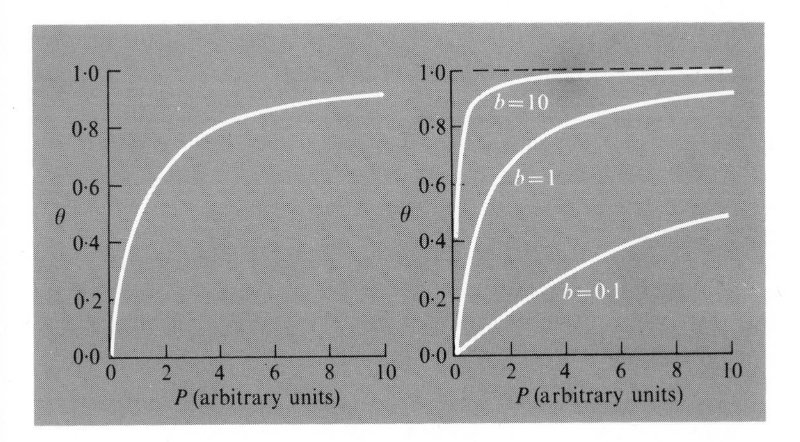

FIG. 3.1. The Langmuir adsorption isotherm.

FIG. 3.2. Langmuir adsorption isotherms with various values of b.

Figure 3.2 shows a number of isotherms with different values of b. We see that the larger the value of b, the greater the surface coverage at a given equilibrium pressure, or the lower the pressure at which a specified coverage is attained. It was explained above that adsorption is almost always exothermic; when it is, application of Le Chatelier's Principle shows that b must decrease with increase of temperature. The isotherms in Fig. 3.2 might equally represent measurements made at different temperatures. Since b is an equilibrium constant, we can apply the van't Hoff isochore and derive the enthalpy change, that is, the heat of adsorption, from its temperature dependence (see Chapter 1). The best procedure is to measure at a series of temperatures the pressures required to give a chosen value of θ: the values of b are then determined from the relation

$$b = \frac{\theta}{P(1-\theta)}.$$

Note that b equals $1/P$ when θ is one-half.

Not all adsorptions obey the Langmuir isotherm. There are several reasons for this, but the most important point is that the assumption concerning the energetic equivalence of all sites is rarely found to be true in practice. More usually the heat of adsorption (which is closely related to the strength of the bond between the adsorbed species and the surface) falls with increasing surface coverage, either linearly or logarithmically. The reasons for this effect have been much discussed, and the most likely explanation seems to be that the first-adsorbed molecules use some of the adsorption potential of atoms neighbouring those on which adsorption actually occurs: molecules arriving later cannot then adsorb so strongly. Other isotherms have been devised to eliminate the assumption of energetic equivalence. The *Temkin isotherm*

$$\theta = k_1 \ln(k_2 bP),$$

where k_1 and k_2 are constants whose values depend on the initial heat of adsorption, supposes a linear decrease of heat of adsorption with coverage. The *Freundlich isotherm*

$$\theta = kP^{1/n},$$

where k and n are constants, the latter being greater than unity, assumes a logarithmic decrease. This isotherm predicts no finite limiting value for θ and is therefore unrealistic; it is however often obeyed for values of θ between 0·2 and 0·8. Care is necessary in deciding which isotherm the observations obey, since, particularly in the region of median coverages, results may seem to fit two or more isotherms equally well.

When two different gases A and B compete for the same sites, and each adsorbs on one site only without dissociation, the fraction of surface uncovered is $1-\theta_A-\theta_B$ (θ_A and θ_B being the fractional coverages of A and B respectively):

application of the procedure described above then gives

$$\theta_A = \frac{b_A P_A}{1 + b_A P_A + b_B P_B}$$

and

$$\theta_B = \frac{b_B P_B}{1 + b_A P_A + b_B P_B},$$

b_A and b_B being the adsorption coefficients for A and B respectively. Thus when the pressures of A and B are equal, the ratio θ_A/θ_B is given by b_A/b_B. The reader should be able to extend this argument to devise a general expression for θ_A when i different adsorbing molecules are present in the system.

One of the best known and most annoying features of heterogeneous catalysts is their ability to be inactivated or 'poisoned' by small amounts of certain substances. A *poison* may enter as an impurity in the reactants and may be either temporary or permanent in nature, depending on whether its effect ceases or not when it is eliminated. It may be generated as a reaction by-product, as for example when carbon deposition occurs during reaction of a hydrocarbon. Metallic catalysts are particularly sensitive to poisons, especially to compounds of sulphur and nitrogen containing lone pairs of electrons which form strong donor bonds to the metal surface: certain elements (particularly mercury and lead) and their ions are also very toxic. A poison is therefore a substance which is much more strongly adsorbed than the reactants, and therefore denies them access to the surface for reaction: its adsorption coefficient is very large, larger relative to those of the reactants when it is a permanent poison than if it is only a temporary one. Note that if b_B is very much greater than b_A, the expression for θ_A reduces to

$$\theta_A = \frac{b_A P_A}{b_B P_B},$$

i.e. it is inversely proportional to P_B.

Adsorption measurements have a variety of uses and applications, but perhaps the most important from our point of view is in determining the *surface areas* of catalysts. Provided the rate of the reaction is not limited by a mass-transfer process (see section 3.4), it will be accurately proportional to the surface area of the active phase of the catalyst. The *activity* of a catalyst is properly expressed as the rate per unit area of surface (usually per m^2), and comparisons between different catalysts should be made on this basis. Measurement of surface areas of catalysts is therefore a matter of great importance. Provided these are made at a temperature well above the boiling point of the gas used, so that a physically adsorbed layer does not build up over the primary chemisorbed layer, then the maximum number of molecules adsorbed (i.e.

that giving monolayer coverage, and corresponding to a value of θ of unity) can be used to estimate the surface area. All we need is a knowledge of the area of surface occupied by each molecule: this comes either from the liquid molar volume or the collision diameter of the gas. Then

$$\text{total surface area} = (\text{number of molecules}) \times (\text{area per molecule}).$$

The application of measurements made on porous solids at temperatures close to the boiling point of the gas will be indicated in the following chapter.

3.2. Kinetics of catalysed reactions: orders of reaction

Measurement of the kinetic parameters of a catalytic reaction is important for a number of reasons. First, knowledge of these, and especially of the orders of reaction with respect to reactants *and products* is essential to (but not always sufficient for) definition of the mechanism of the reaction, and it is only through some understanding of reaction mechanism that catalyst optimization may be carried out on a rational and scientific basis. Second, the best design of the catalytic reactor, including the size and shape of the catalyst bed, depends critically on information concerning the reaction orders. Third, the magnitude of the activation energy tells us how temperature will affect the rate of the reaction (see section 3.3).

These statements assume that we have a simple, clean conversion of reactants into a single desired product, and that the kinetics relate to a stoichiometric equation which can be defined without ambiguity. This situation is rarely encountered in practice. A much more common situation is that where unwanted by-products are formed simultaneously either by a parallel reaction, symbolized as

$$A + B \begin{matrix} \nearrow C \\ \searrow X \end{matrix}$$

or by a sequential reaction of the desired product, viz.

$$A + B \rightarrow C \rightarrow X:$$

here C is the desired product, and X the undesired. In such cases we need to know the kinetic parameters of each contributing reaction. The undesired by-product may remain on the surface and constitute a catalyst poison, in which case knowledge of its rate of formation is vital. Description of the rate equations for these important but rather complicated systems cannot be attempted here.

We may however initiate a short treatment of the kinetics of simple catalytic reactions on the basis of the Langmuir isotherm, notwithstanding its limitations. Let us consider first a unimolecular conversion of a molecule A, adsorbed without dissociation, into a product C which is not adsorbed at all. The rate of removal of A from the gas phase will depend only on the concentration of adsorbed A, that is, on its surface coverage; thus

$$-\frac{dP_A}{dt} = k\theta_A = \frac{kb_A P_A}{1+b_A P_A},$$

k being a rate coefficient. There are two important limiting conditions:

(1) when either b_A or P_A is so small that $b_A P_A$ is much less than unity, θ_A approximates to $b_A P_A$ and the reaction is therefore first order in A: under these conditions θ_A is of course low;

(2) when either b_A or P_A is so large that $b_A P_A$ is much greater than unity, θ_A becomes almost independent of P_A and the reaction is zero order in A: this occurs when θ_A is in the region of unity.

When neither of these approximations applies, the full form of the rate expression must be used, and the order in A (at least over a limited pressure range) appears to have a positive fractional value between zero and unity. If it were feasible to examine the order of such a reaction over an extended pressure range, we should thus see first-order behaviour at low pressures changing through intermediate values to zero order at high pressures.

Many catalytic reactions are bimolecular, and the expressions given above for the surface coverages given by two gases admixed may be used to discuss the kinetics of such reactions. Thus the rate of the process

$$A + B \rightarrow C$$

is given by

$$\frac{+dP_C}{dt} = k\theta_A\theta_B = \frac{kb_A P_A b_B P_B}{(1+b_A P_A + b_B P_B)^2}$$

provided (i) the molecules A and B are adsorbed on separate sites and without dissociation, (ii) the slow step is the reaction between the two adsorbed molecules, and (iii) the product C is not adsorbed. There are now several limiting cases to consider, of which we will take only two.

(1) If A and B are both weakly adsorbed, i.e. if b_A and b_B are both very much less than unity, the rate expression reduces to

$$\frac{+dP_C}{dt} = k' P_A P_B,$$

where k' equals $kb_A b_B$: the reaction is first order in each reactant, and is overall second order.

(2) If A is weakly adsorbed and B is strongly adsorbed, then $b_A \ll 1 \ll b_B$, and the rate expression reduces to

$$\frac{+dP_C}{dt} = \frac{k'' P_A}{P_B},$$

where k'' is kb_A/b_B: the reaction is first order in A, and of minus first order in B.

Returning to the full expression, the rate is a maximum when θ_A equals θ_B, that is, when b_AP_A equals b_BP_B. If the partial pressure of A is changed while that of B remains fixed, then the ratio of P_A/P_B at the maximum rate gives the value of b_A/b_B. Numerous other rate expressions have been developed to treat more complicated situations, but enough has probably been said to indicate the approach.

In this elementary treatment of the kinetics of heterogeneous reactions, we have assumed that, where two reactants are present, both must be adsorbed side by side for reaction to occur, and that reactions such as

$$A_g + B_a \rightarrow C_g$$

or

$$A_a + B_g \rightarrow C_g$$

(the subscripts signifying adsorbed and gaseous) are unimportant. Indeed in some systems there is definite evidence that such processes participate. For the first alternative the rate expression becomes

$$\frac{+dP_C}{dt} = kP_A\theta_B = \frac{kP_Ab_BP_B}{1+b_BP_B}.$$

When b_B or P_B is large, the rate becomes first order in A and *zero* order in B: if A were adsorbed only weakly, then the orders would be first in A and minus first in B (see above).

It must be clearly understood that knowledge of a reaction's kinetics does not of itself define the mechanism of the reaction, since there may be many fast elementary steps intervening between reactants and products; but since they are not rate-determining, they do not appear in the kinetic equation, or rate expression. *Orders of reaction give information about the slowest step only; knowledge of them is necessary but not sufficient to describe the mechanism.* We shall later encounter examples of the wealth of mechanistic detail unrevealed by a simple kinetic study.

3.3. Kinetics of catalysed processes: temperature effect

There is one important difference between homogeneous and heterogeneous reactions with regard to effect of temperature. In the case of a homogeneous unimolecular conversion of A, we would write

$$\frac{-dC_A}{dt} = kC_A,$$

C_A being the concentration of A: k is of course easily determinable by experiment and its temperature dependence leads to the activation energy (see

Chapter 1). In the case of the corresponding heterogeneous reaction where the product is not adsorbed, we saw above that

$$\frac{-\mathrm{d}P_{\mathrm{A}}}{\mathrm{d}t} = k\theta_{\mathrm{A}}.$$

The effect of temperature upon *rate* is therefore the product of its effects on k and θ_{A} separately, but now the rate coefficient cannot be unequivocally estimated because the surface coverage is not directly measurable. We must therefore enquire into the effect of temperature on this latter quantity.

Once again we have two limiting conditions to consider.

(1) If the adsorption coefficient b_{A} is sufficiently large that the surface coverage θ_{A} remains effectively unity throughout the temperature range examined, then the whole of the temperature effect is upon the rate constant and we can derive a true activation energy E_{true} from the Arrhenius equation

$$k = A\exp(-E_{\mathrm{true}}/RT).$$

(2) If however b_{A} is small, and θ_{A} is close to zero, we have from the van't Hoff isochore

$$\frac{\mathrm{d}\ln b_{\mathrm{A}}}{\mathrm{d}T} = \frac{\Delta H_{\mathrm{a}}^{\ominus}}{RT^2}$$

or

$$b_{\mathrm{A}} = C\exp(-\Delta H_{\mathrm{a}}^{\ominus}/RT)$$

where C is an integration constant and $-\Delta H_{\mathrm{a}}^{\ominus}$ is the standard molar heat of adsorption of A. Remembering that θ_{A} equals $b_{\mathrm{A}}P_{\mathrm{A}}$ under these conditions, the effect of temperature on the rate is given by the equation

$$\frac{-\mathrm{d}P_{\mathrm{A}}}{\mathrm{d}t} = P_{\mathrm{A}}AC\exp[(-E_{\mathrm{true}}-\Delta H_{\mathrm{a}})/RT]$$

$$= P_{\mathrm{A}}AC\exp(-E_{\mathrm{app}}/RT)$$

where E_{app} is the observed or apparent activation energy. Thus

$$E_{\mathrm{true}} = E_{\mathrm{app}} - \Delta H_{\mathrm{a}}.$$

However since adsorption is always exothermic, ΔH_{a} is negative from the system's point of view and $-\Delta H_{\mathrm{a}}$ has a positive value. The true activation energy is thus obtained by adding the heat of adsorption as a positive quantity to the apparent activation energy.

The situation which might be observed if a reaction of this kind were studied over a wide range of temperature is shown in Fig. 3.3. The dotted line indicates the rates which would have been observed had the surface coverage not fallen below unity. It is important to distinguish this cause of apparent

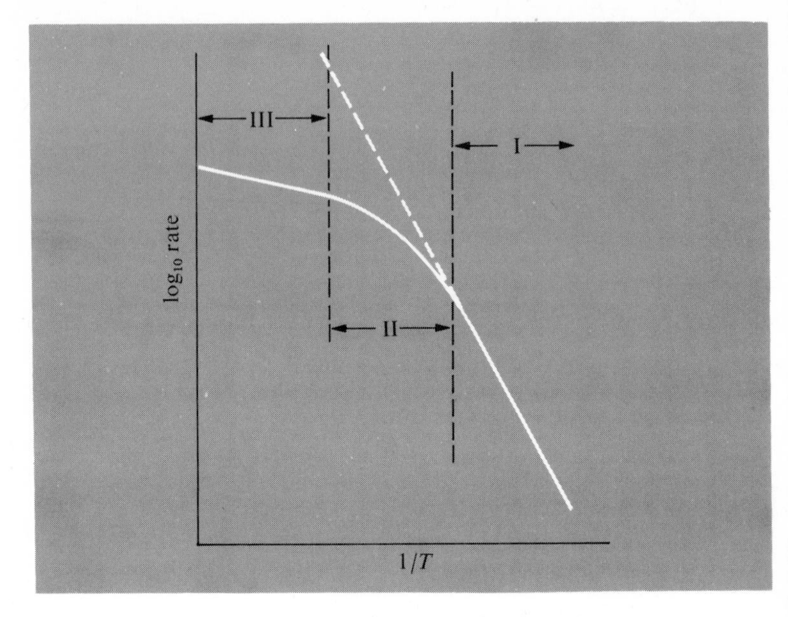

FIG. 3.3. Arrhenius plot for a catalysed reaction.

	I	II	III
Surface coverage, θ	1	$1 > \theta > 0$	~ 0
Order, n	0	$1 > n > 0$	1
Slope × 2·3 R gives	E_{true}	—	E_{app}

change in activation energy from a quite different cause (see section 3.4) which has a very similar effect.

3.4. Diffusion limitation

So far we have tacitly assumed that the reaction between adsorbed molecules, or between adsorbed and gaseous molecules, determines the overall rate of reaction: unfortunately this is not always the case. The total process may be broken down into the following five basic steps, any one of which can be rate determining.

(1) Transport of reactants to the catalyst.
(2) Adsorption of the reactants on the catalyst.
(3) Reaction on the catalyst involving one or more adsorbed reactants.
(4) Desorption of products from the catalyst.
(5) Transport of products away from the catalyst.

Steps 2, 3, and 4 are chemical in nature, and may be regarded as jointly constituting the catalytic reaction: assignment of the slowest step in a particular case is, as we have mentioned above, not always self-evident from the observed kinetics. Steps 1 and 5 on the other hand involve no chemical change. Step 1 is the physical process whereby the reactants are brought through the gaseous or liquid phase surrounding the solid catalyst to the active sites on the latter's surface. This is a *diffusion process*, and the phenomenon is called *mass transport* or *mass transfer*. Step 5 is the corresponding process for getting products away from the surface. When either of these is slower than the catalytic rate itself, the rate is determined by the rate of arrival of reactants (or of removal of products): we then say the rate is *diffusion limited* or *mass-transport limited*.

It is most important in practice to be able to recognize when diffusion limitation is operative, since its occurrence signifies that the catalyst is being used to less than its maximum capacity. There are some occasions when it is preferable that the reaction be diffusion limited, as when for example the isolation of an intermediate product in high yield is desired (see section 3.2), but more often it is better to eliminate it to make better use of the catalyst. Diffusion limitation is recognized by the following characteristics.

(1) The rate is proportional to the catalyst weight (or to the concentration of the active component) raised to a power less than unity, which in the limit may be zero.
(2) The rate is increased by improving the movement of the gas or liquid with respect to the catalyst.
(3) The temperature coefficient is low, and the apparent activation energy may be as low as 10–15 kJ mol^{-1}: gaseous diffusion processes do not in fact obey the Arrhenius equation, their rates being proportional to $T^{\frac{1}{2}}$.

Reactions whose rate is governed by a truly chemical step on the other hand show the following features.

(1) The rate is accurately proportional to catalyst weight or the concentration of the active component.
(2) The rate is unaffected by better agitation.
(3) The apparent activation is usually in excess of 25 kJ mol^{-1}.

Reactions where a liquid phase is present are far more likely to become diffusion limited than those where only gaseous reactants participate. This is because *diffusion coefficients* are much lower in liquids than in gases. Diffusion can only occur in the presence of a concentration gradient, and the diffusion coefficient D is defined by *Fick's first law of diffusion* (1855) in the following way:

$$\text{rate of diffusion} = \frac{\mathrm{d}c}{\mathrm{d}t} = D \, . \frac{\mathrm{d}c}{\mathrm{d}x}$$

where $\mathrm{d}c/\mathrm{d}x$ is the concentration gradient. Diffusion problems are especially encountered where both gaseous and liquid reactants are present, for example, in the catalytic hydrogenation of liquid benzene. Because of the low solubility of hydrogen in benzene, the concentration gradient between the catalyst surface and the bulk liquid can never be large, and as seen from Fick's first law this determines the diffusion rate. For this reason it is always necessary to agitate such systems vigorously. This has the effect of (i) extending the liquid-gas interface, which increases the rate of hydrogen dissolution and thus maximizes the concentration of dissolved hydrogen; and (ii) minimizing the distance which a hydrogen molecule has to diffuse through the liquid phase in order to find the catalyst surface. The condition of diffusion limitation is denoted by the presence of a concentration gradient close to the surface (see Fig. 3.4). An additional way of speeding up a two-phase reaction system is to use superatmospheric pressures of the gas since, according to Henry's Law, the concentration of dissolved gas will be proportional to its pressure. Thus even working at 5 or 10 atm (0·5–1 MPa) will lead to a significant improvement over working at 1 atm, but one which is not sustained proportionally at very high pressures. However pressures of the order of 100 atm (10 MPa) are not infrequently used.

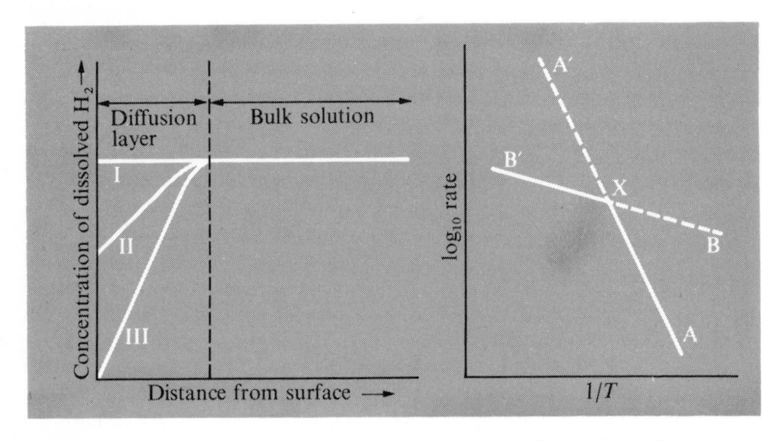

FIG. 3.4. Concentration gradients in the neighbourhood of a surface.
I No diffusion limitation
II Partial diffusion limitation
III Complete diffusion limitation.

FIG. 3.5. Arrhenius plot showing the onset of diffusion limitation
AA′ surface reaction is rate-limiting
BB′ reaction is diffusion-limited
Observations follow AXB′ since the slower of the two processes controls the rate.

We must finally note an important consequence of the small temperature coefficient for diffusion. A reaction may be free from diffusion limitation at low temperatures, but may suffer from it at higher temperatures (see Fig. 3.5).

The effect is similar to that depicted in Fig. 3.3 (even to the kinetic consequences), but the cause is quite different. In general anything which reduces the rate of the chemical step (lower temperature, less catalyst, or less of the active component) makes diffusion limitation less likely.

QUESTIONS

3.1. Show by means of a potential-energy diagram how, in a system where the heat liberated on chemisorption decreases with increasing surface coverage, the adsorption may have no energy of activation at low coverages although one may develop as the adsorption proceeds.

3.2. Construct a set of numbers obeying a Langmuir adsorption isotherm and plot the results: this is a graph of the equation $y = x/(1+x)$, and you plot in effect y versus x (see Fig. 3.1).
(a) Do this for values of b of 0·1, 1, and 10 (see Fig. 3.2).
(b) There are at least three ways of changing the expression for the isotherm into a linear form. Test your set of numbers against all the linear forms you can find. Which do you think is the best, and why?
(c) Apply the Temkin and Freundlich isotherms to your sets of numbers. Is it clear that the form of these equations differs from that of the Langmuir isotherm? Over what pressure range might the isotherms be confused?

3.3. Modify the Langmuir adsorption isotherm to allow for fragmentation of the molecule into (a) two and (b) n parts on adsorption.

3.4. Write down a general expression for θ_A where A is adsorbing on a surface in the presence of a total of i different species, all obeying the Langmuir isotherm.

3.5. Two different molecules, A and B, are each adsorbed in obedience to the Langmuir isotherm, and reaction occurs between two adsorbed molecules to form a non-adsorbing product. Show graphically how
(a) the rate varies with the ratio of θ_A to θ_B;
(b) the rate varies with P_A if P_B is kept constant, using ratios of b_A/b_B of 0·1, 1, and 10.

4. Heterogeneous catalysts: preparation, structure, and use

4.1. Catalysts for basic research

WE have said little thus far about how catalysts are made, how their physical structure is examined, or how they are employed in practice, either in the laboratory or in industry. The physical form in which a catalyst is used depends upon the purpose of its use: thus very often the forms employed in basic research are quite different from those which may be used in large scale practice. This is especially so with metal catalysts. For fundamental work it is extremely important that the catalyst should have a known chemical composition, and that its surface should be clean or readily cleansed, or at least that the concentration and nature of impurities on the surface be ascertainable. For this reason, basic work on catalysis by metals frequently employs wires, which may be cleansed by heating them electrically; foils and single crystals, cleaned by ion bombardment; or thin films prepared by condensation of atoms evaporating from a wire heated close to the melting point in high vacuum. Alloys can also be made in this way. Although all these forms of metals have rather low surface areas, they have the great advantage of being susceptible to examination by such powerful physical techniques as LEED (low-energy electron diffraction), Auger spectroscopy, and ESCA (electron spectroscopy for chemical analysis). While it might be thought that these metallic forms are quite dissimilar to those of practical metal catalysts, it is now possible to cut single crystals at a slight angle to a low index plane, thus making a 'stepped' surface which replicates some of their characteristics. Since little more will be said here concerning the direct physical examination of surfaces or of species adsorbed thereon, it should be pointed out that there are grave difficulties in deciding whether adsorbed species seen on these 'clean' surfaces are actually those which participate in the catalytic act. The strong possibility remains that catalytically reactive species are only a small part of the total, and hence not easily visible.

4.2. Technical catalysts—general considerations

The criteria for an industrially successful catalyst are quite different, and more stringent. The two most important considerations are activity and durability; suitability for basic study is not a factor, and this has retarded progress in our understanding of how such technical catalysts really operate. Let us consider further the two prime requirements. First, the catalyst must be able to effect the desired reaction at an acceptable rate under conditions of temperature and pressure that are practicable. Chemical technology has advanced to the point where temperatures as high as 1600 K and pressures up to 350 atm (35 MPa) present no difficulties, and thermodynamic considerations sometimes necessitate their use to attain reasonable equilibrium yields of products (see section 1.1). If however good yields can be got at low temperatures and pressures, then there is every incentive to find a catalyst that will operate under the mildest possible conditions, since the use of extreme conditions is very costly. It is concurrently important that side reactions are minimal, especially those leading to poisoning or deactivation through carbon deposition.

Second, the catalyst must be able to sustain the desired reaction over prolonged periods: in some processes, a catalyst life of several years is not uncommon, and is economically necessary. Clearly the longer it lasts, the smaller will be the contribution which its initial cost makes to the overall cost of the process. Initial cost is rarely of over-riding importance: it is usually cheaper in the long run to use an expensive catalyst that will last a long time than a cheap one that has to be replaced frequently. The chief causes of deterioration in use are (i) reversible poisoning due to impurities in the reactants or to side reactions, and (ii) irreversible physical changes including loss of surface area (sintering) or mechanical failure. Reversible poisoning may with luck be rectified by simple treatment, such as oxidation or washing, without removing the catalyst from the reactor. To guard against physical changes, careful attention has to be given to the strength of the catalyst.

4.3. Catalytic reactors and physical forms of catalysts

The preferred physical form for a catalyst is determined entirely by the manner in which it is to be used. Its particle size especially is fixed, at least within broad limits, by the type of reactor to be employed. Table 4.1 lists the

TABLE 4.1

Physical forms of catalyst suitable for use with various reaction systems and reactor types

Reaction system	Reactor types	Catalyst forms
Gases only	fixed-bed	coarse particles or monolithic structure
	fluidized-bed	fine particles
Gas + liquid	batch reactor*	fine particles
	bubble column reactor	fine particles
	continuous stirred-tank reactor*	coarse particles
	trickle column reactor*	coarse particles or monolithic structure

* These reactor types are also applicable to systems where only a liquid reactant is to be used.

chief reactor types, and specifies the general form of catalyst compatible with each. The terms 'coarse' and 'fine' are used very qualitatively, a rough criterion being whether or not one can distinguish separate particles with the naked eye.

When the reactants are solely gases or vapours, there are only two possible basic types of reactor which can be used. A *fixed-bed reactor* is one in which a tube is packed with 'coarse' catalyst particles through which the reactants flow. As a result of the obstruction to gas flow by the particles, a pressure drop occurs across the bed, and a positive pressure has to be applied at the inlet to secure an adequate flow rate. The size of the pressure drop increases with increasing flow rate and bed length, and with decreasing particle size. Many different shapes and sizes of catalyst particles are used in fixed-bed reactors, and a great deal of skill and experience goes into choosing the most appropriate for a given set of conditions. Some of the more common forms are sketched in Fig. 4.1; dimensions are usually between 2 mm and 2 cm.

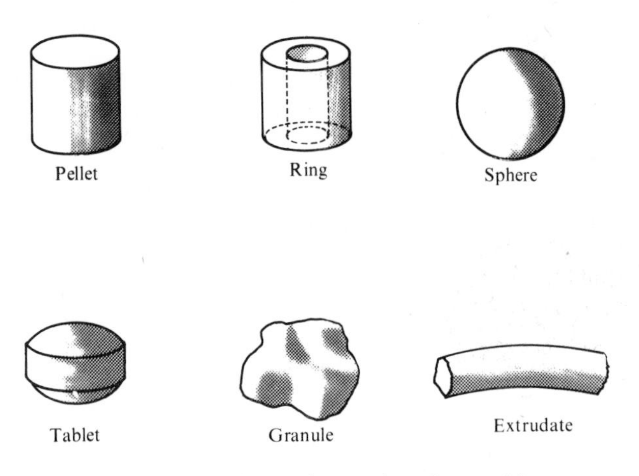

FIG. 4.1. Various forms of 'coarse' catalyst particles.

The average time a reactant molecule remains in the catalyst bed is termed the *contact time*: this is given by the free volume in the bed including the pore volume divided by the flow rate of the gas. The reciprocal of the contact time is the *space velocity* which is the number of volumes of gas flowing through the free volume of the bed in unit time.

When as in the case of ammonia oxidation it is desired to use a pure metal catalyst at high temperature and a very short contact time, a bed of 20 to 30 finely woven metal gauzes is used (see section 10.3); this is another catalyst form suited to use in a fixed-bed reactor.

Since many catalysed reactions are strongly exothermic, fixed-bed reactors must contain a facility for removing the unwanted heat of reaction. This may be done in one of two ways (see Fig. 4.2). A *multitubular reactor* contains a large number of reactor tubes, typically 2.5 cm in diameter, with a cooling gas or liquid flowing between them. Alternatively the bed may be split into sections, with provision for cooling the gas in the spaces between.

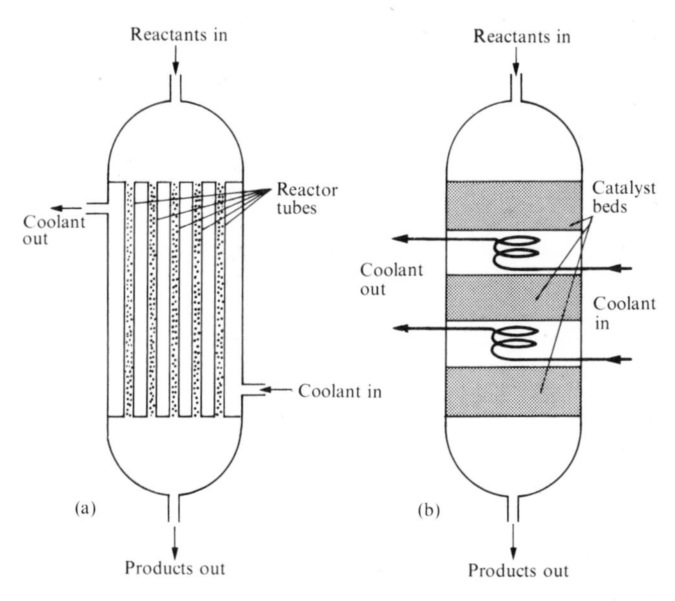

FIG. 4.2. Diagrammatic representation of fixed-bed catalytic reactors: (a) multitubular reactor, (b) sectioned-bed reactor.

A further important and quite different catalyst form is finding application especially in air-pollution control (see Chapter 11). This is the so-called *monolithic structure*, which consists of a block of ceramic material (α-alumina or mullite) through which run fine parallel channels. The structures can be made in many different ways, each of which results in a characteristic channel shape: some of these are indicated in Fig. 4.3. The blocks are obtainable in a range of shapes and sizes, and a single block when inserted in a container therefore behaves as a fixed-bed reactor. Because of the low porosity of the material from which these structures are made, it is usually necessary to attach a thin layer of a more porous substance (e.g. γ-alumina) to which the catalytically active phase can adhere. Such monolithic structures have several clear advantages over pellets or granules packed loosely in a tube. First, the pressure

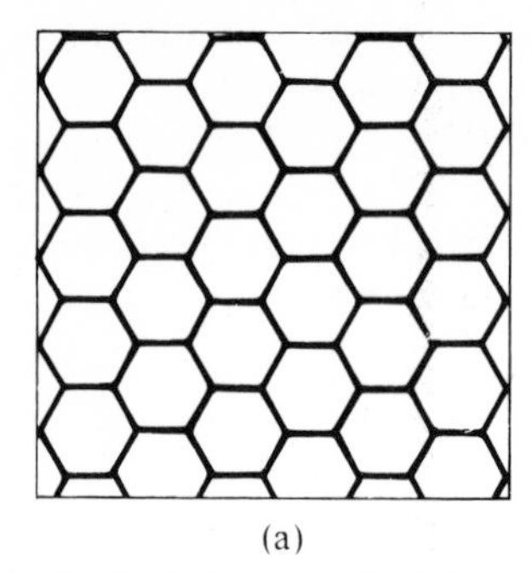

(a)

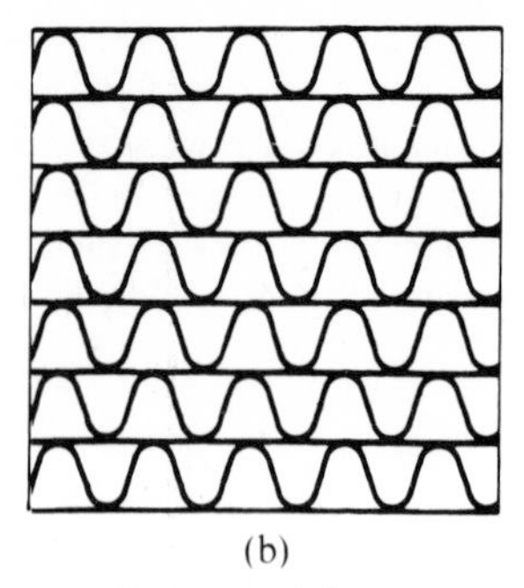

(b)

FIG. 4.3. Typical cross-sections of monolithic supports: (a) honeycomb, (b) corrugated.

drop through the bed is less for equal quantities of catalyst, thus permitting higher space velocities to be achieved; secondly, there is no *attrition* of the catalyst due to particles rubbing against each other and forming fine dust. Their ability to withstand thermal shock is however in some cases limited.

The second basic kind of reactor for gaseous reactions is the *fluidized bed*. Here the catalyst consists of fine particles, and when the gas flow upwards through a bed of such powder attains a critical velocity the bed appears to 'boil': it expands significantly, and the particles are in continuous motion. In this state the bed is said to be fluidized. This behaviour gives certain advantages over fixed beds: for example, heat transfer characteristics are much better, and the pressure drop increases far less quickly with increasing flow rate (see Fig. 4.4).

Table 4.1 sets out the physical forms of a catalyst compatible with various reactor types and reaction systems, and shows that there are a larger number of kinds of reactor suited for use with liquid plus gaseous reactants. Probably the most commonly used is the *batch reactor*. In this the catalyst as a fine powder is suspended in the liquid reactant or a solution thereof and is maintained in suspension by shaking or stirring, or by the movement of gas through the liquid. This agitation is necessary to provide the good three-phase contact on which efficient reaction depends (see sections 3.4 and 6.3). If the reaction is conducted under superatmospheric pressure, as is often the case, the vessel used is termed an *autoclave*. A slight variant of this is the *bubble-column reactor* in which the catalyst, again dispersed in a liquid, is placed in a long tube and is kept in suspension by a stream of fine bubbles of the gas. This device is particularly suitable for basic kinetic studies, as the liquid may be analysed at various points along the tube and the kinetic equation thus established.

To perform a catalytic reaction where the reactants are gaseous and liquid in a continuous manner is not easy, but there are two ways in which it may be done. If a suitable 'coarse' catalyst is packed into a column, the liquid reactant

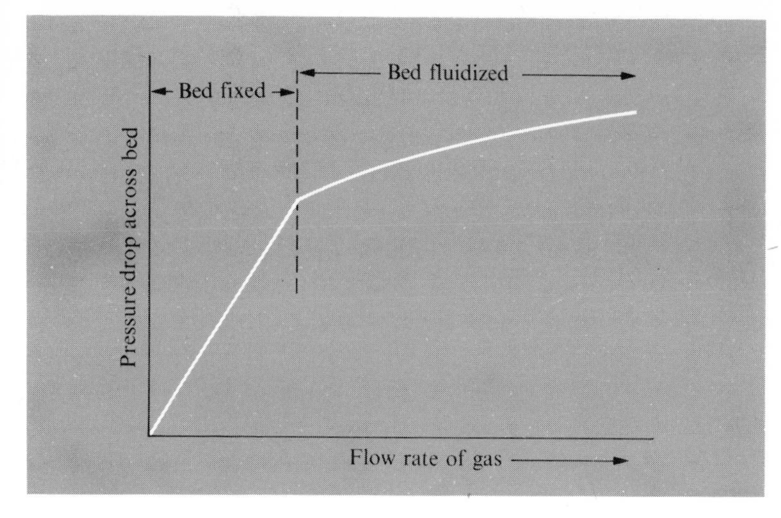

FIG. 4.4. Pressure drop versus flow rate for fixed and fluidized beds. Fluidization commences at the flow rate indicated by the broken line.

or a solution can be trickled down over the bed, and the gaseous reactant caused to flow either up the bed (counter-currently) or down the bed (concurrently) (see Fig. 4.5). The product, or product–reactant mixture, is then collected at the bottom of the column. This device is known as a *trickle-column reactor*. Alternatively a 'coarse' catalyst can be held in a wire-mesh basket which is rotated in the liquid, while the gas is bubbled in below it. A continuous slow flow of liquid reactant or solution into the vessel is balanced by abstraction of the liquid at another point: by adjusting these flow-rates, the reactant in the liquid resides in the vessel long enough on average to achieve the desired degree of conversion. This system is termed a *continuous stirred tank reactor* (Fig. 4.6).

4.4. Composition and structure of catalysts

It has been noted above (section 3.1) that the rate of a catalytic reaction should be proportional to the surface area of the catalyst, provided that transfer of reactants to or of products from the surface is not the slow step. It is therefore normally desirable to get as much surface area as possible into a given volume. There is a limit to the surface area which may be achieved simply by progressive subdivision of the catalyst. For a material of density 3 g cm^{-3}, in the form of uniform microspheres 50 nm in diameter, the surface area is only 40 m^2 g^{-1}. This is not a large surface area for a catalyst, and the difficulty of using so fine a powder would be considerable. The answer lies in making

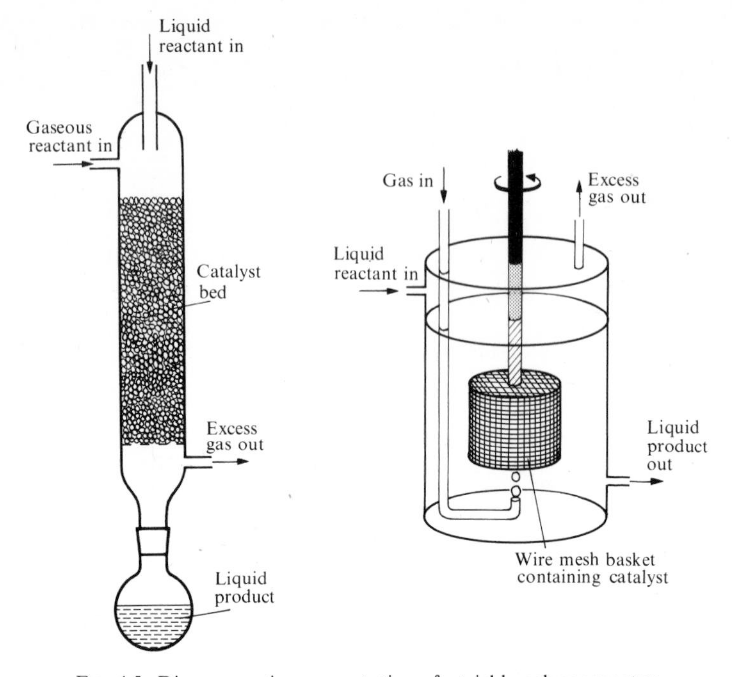

FIG. 4.5. Diagrammatic representation of a trickle-column reactor.
FIG. 4.6. Diagrammatic representation of a continuous stirred tank reactor.

the material *porous*, by forming a large number of very fine channels or *pores* through the particles. For non-intersecting uniform cylindrical pores, the relation between the internal surface area S, the pore volume V_p and the radius r of the pores is given by

$$S = 2V_p/r.$$

Thus for a substance having pores of 2 nm radius and 1 $cm^3\ g^{-1}$ pore volume, the surface area would be 1000 $m^2\ g^{-1}$. Such values are commonly encountered in practice, and may moreover be found with 'coarse' particles which are easily handled. Of course, except in the special case of zeolites (see section 2.9), pores are never uniform in size or shape. They are divided into three groups on the basis of their size: pores above 100 nm in width are *macropores*; those between 2 and 100 nm are *mesopores*; and those below 2 nm in width are *micropores*. Pores in the first two groups are usually spaces between primary particles, but micropores are usually fissures of near-atomic dimensions within them. Information on the porosity of solids comes from examining the physical adsorption of gases near their boiling points, and by application of some other techniques (section 4.5).

There are however some circumstances under which it is undesirable to use highly porous catalysts: this is especially so when we wish to obtain an intermediate product in high yield (see section 3.2). Such a product formed in the depth of a pore would stand a very high chance of suffering further reaction on the way out. For this reason the use of low-surface-area catalysts is much preferred in these cases.

There are many elements and compounds which, though catalytically active, cannot be readily obtained as porous solids of high surface area. A widely used procedure to overcome this difficulty is to distribute the active component over the surface (including the internal surface) of particles of high porosity: γ-alumina, silica gel, and activated charcoal are often used for this purpose, and when so used they are referred to as *catalyst supports*. These and many other substances may be used in either 'fine' or 'coarse' forms, and in most of the shapes described above. In a supported catalyst, the active phase usually constitutes 0·1 to 20 per cent by weight of the total catalyst, and is normally in the form of very small crystallites (1 to 50 nm in diameter). The support itself is generally without catalytic activity in the desired reaction, but there are some important exceptions to this generalization (see Chapter 7). The support can also modify the catalytic properties of the active phase in a useful way. The preparation and characterization of supported metal catalysts is considered further below (section 4.6). Examples of supported catalysts are chromium trioxide on silica for olefin polymerization, and palladium on charcoal powder for hydrogenation.

In some instances the active phase can constitute the great part of the total catalyst, but small amounts of other materials (typically 5 to 25 per cent by weight of the whole) may be present: these are called *catalyst promoters*. Their role is complex, but one clearly distinguishable function is to inhibit growth in size of the small crystals of the active material during use. This growth if not stopped would lead to loss of surface area and hence of activity. Examples of promoted catalysts are iron catalysts for ammonia synthesis (see Chapter 10) and nickel–alumina catalysts for the steam reforming of hydrocarbons (see Chapter 9).

4.5. Characterization of porous substances

Our concern here is with the internal surface area of a porous solid, and with the nature of the pores giving rise to this area. With most porous solids the external surface is a negligible fraction of the total, but if relevant it may readily be estimated by measuring the permeability of gas through a bed of the solid, or by sedimentation. The parameters of interest are the internal surface area and the pore volume, from which an average pore radius may be obtained; and the pore size distribution. In addition it is useful to know something about the shape of the pores.

The pore volume may be crudely estimated by a simple titration with a liquid, although more refined methods are of course available. Knowledge of the pore size distribution is obtainable by mercury porosimetry. In this technique, mercury is forced under pressure into the pores, and the greater the pressure, the smaller the pores to which the mercury obtains access. By sensitively measuring the change in volume of the solid plus the mercury with increasing pressure, a picture of the pore size distribution is obtained. A pressure of 70 MPa is needed to get mercury into pores of 10 nm radius, and this is the practical limit; however pores up to 10^4 nm in radius can be sensed.

A great deal of very useful information is derived from measurement of the physical adsorption of gases on porous solids. We have seen above (section 3.1) that the monolayer capacity of a non-porous solid, measured by chemisorption, or by physical adsorption well above the boiling point of the adsorbing gas, can be easily translated into a surface area. However with porous solids, and using temperatures close to the boiling point of the gas, so that multilayer adsorption occurs, several forms of isotherm besides the Langmuir type have been observed. These have been classified into five types, as shown in Fig. 4.7.

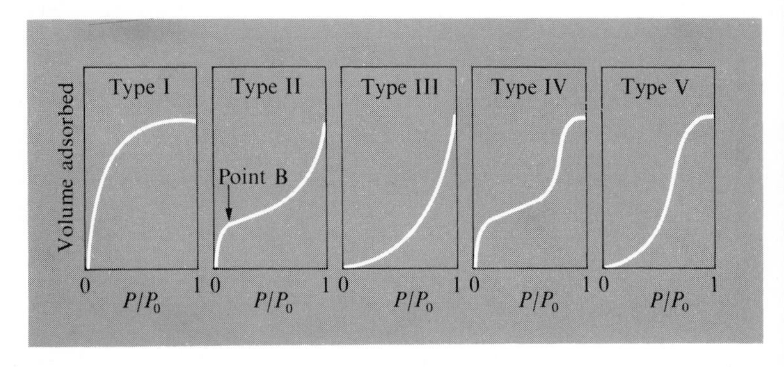

FIG. 4.7. Classification of isotherms according to the BET theory.

The first (Type I) is of the form of the Langmuir isotherm, but it is given by microporous solids, including zeolites, and the plateau probably represents the complete filling of very small pores by the condensed gas. The Type II isotherm is commonly encountered, and application of the Brunauer, Emmett, and Teller (BET) equation in the form

$$\frac{P}{x(P_0-P)} = \frac{1}{x_m C} + \frac{C-1}{x_m C} \cdot \frac{P}{P_0}$$

permits the monolayer capacity x_m to be derived from both the slope and the intercept. We cannot unfortunately derive this equation here. P_0 is the saturated vapour pressure of the adsorbing gas at the temperature used, and C is given by

$$C = \exp[(H_a - H_l)/RT],$$

H_a being the heat of adsorption in the first layer, and H_l the heat liberated on forming the second and subsequent layers, this being equated to the heat of liquefaction of the gas. The monolayer capacity has also been equated with the adsorption at 'point B' (see Fig. 4.7).

Isotherms of Types III and V are of little interest since their theoretical basis is not well understood, but Type IV is of great interest. This isotherm usually shows a *hysteresis loop* (see Fig. 4.8), that is, the isotherm does not follow the

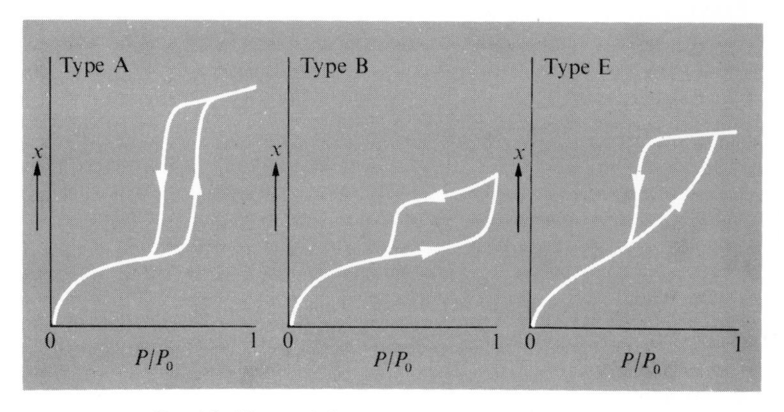

FIG. 4.8. Hysteresis loops on Type IV isotherms.

same path in desorption as it does in adsorption. The reason for this is that evaporation of condensed gas in fine pores does not occur as easily as its condensation; this is because a molecule evaporating from a highly curved meniscus has a higher probability of recondensing than one evaporating from a plane surface. This effect was discussed many years ago by Lord Kelvin who devised the following equation to describe the effect:

$$\ln(P/P_0) = -\frac{2V\gamma}{rRT}\cos\phi,$$

where V is the molar volume of the liquid, γ its surface tension, r the pore radius and ϕ the contact angle (usually taken to be zero): the other symbols have their usual significance. The relative pressure at which condensation will occur in a pore of a given size can thus be determined, and the isotherm then used to obtain a pore size distribution.

Hysteresis loops vary greatly in shape, and these too have been classified: only Types A, B, and E (see Fig. 4.8) need concern us. The Type A loop is given by open-ended cylindrical pores, Type B by open slit-shaped capillaries, and Type E by open or closed pores of variable radius.

4.6. Supported metal catalysts

It is particularly convenient to use metals in the form of supported catalysts. As we saw in Chapter 2, the noble Group VIII metals are frequently extremely active catalysts, but their high cost demands they be used as very small particles having a large surface-to-volume ratio. Although such small particles can be easily made, for example, as colloidal dispersions, they are then not very convenient to use. It is very easy to make fine metal particles on a support. A typical laboratory preparation would be conducted in the following way. Small particles of silica gel in an evaporating dish are wetted with an aqueous solution of chloroplatinic acid containing the desired weight of platinum. The water is removed, first by heating with a steam-bath, and later in a hot-air oven. The dried material is then reduced in flowing hydrogen at 470 K until no more hydrogen chloride is formed. The resulting platinum/silica catalyst will contain metal particles between 1 and 10 nm in diameter. This method is known as an *impregnation method*; of course, some practical detail has been omitted, and the procedure is susceptible to very many variations.

The advantages of using this kind of catalyst are the following:

(1) The product is easily and safely handled, and, if used in a liquid medium, may be recovered by filtration.
(2) The surface-to-volume ratio of the metal is high.
(3) The metal particles do not sinter rapidly, since they are not in close contact.
(4) Small losses can be tolerated if the metal content is low.

The physical characterization of supported metal catalysts is a challenging problem, to whose solution a great many analytical methods have been applied. It is first of all important to know the surface area of the metallic phase, since this is what will determine the effectiveness of the catalyst. This may be estimated by measuring the monolayer capacity of a gas which, under carefully selected conditions, is selectively chemisorbed by the metal; hydrogen, carbon monoxide, and oxygen have been widely used. Assuming the ratio of chemisorbed hydrogen atoms to surface metal atoms to be unity, and knowing the metal content of the catalyst, an estimate can then be made of the *average particle size* of the metal. Some uncertainty still surrounds the stoichiometry of chemisorption, especially for very small particles, and care is necessary in selecting the gas to be used: carbon monoxide cannot be used with nickel because nickel carbonyl might be formed, and hydrogen chemisorption on palladium is differentiated from its solution into the interior only by careful observations.

This procedure cannot of course reveal the breadth of the distribution of particle sizes. This is also useful to know, as it may provide useful clues to the way in which the particles are formed. The most direct way of observing the size distribution is by *transmission electron microscopy*: this can show particles as small as 1 nm, but the method requires skill and experience, and the instrumentation is expensive. An electron micrograph of a supported metal catalyst is shown in Fig. 4.9. If the size of a sufficient number of particles is measured, a meaningful size distribution results; this is often surprisingly

FIG. 4.9 Electron micrograph of a silica-supported Pd–Au alloy catalyst (4% Au, 1% Pd by weight).

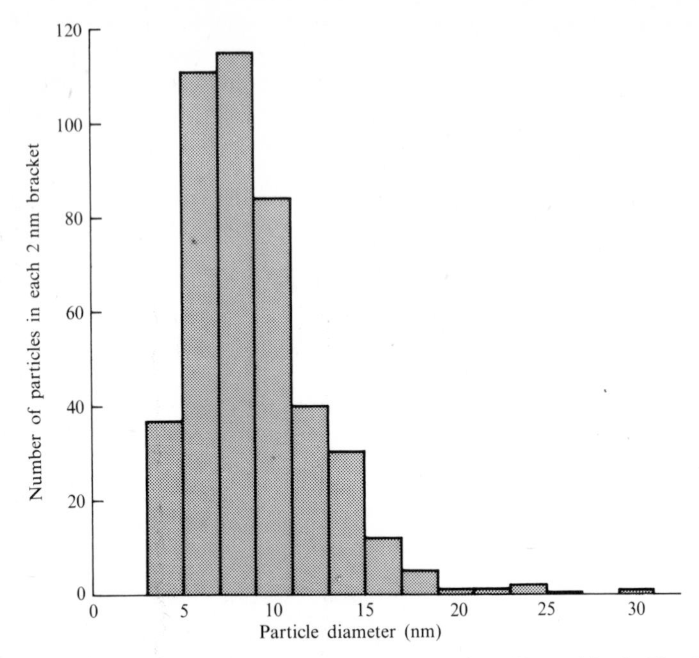

FIG. 4.10 Particle size distribution derived by measuring 439 particles in Fig. 4.9.

wide (see Fig. 4.10). Very small clusters containing only five atoms (of rhodium on silica) have been seen in exceptionally careful work.

One further method must be mentioned, because of its frequent use. X-ray diffraction lines shown by particles less than about 50 nm in size are broadened, and the degree of broadening can be employed to give the average particle size. The required relation is given by the Scherrer equation:

$$B_c = K_1 \lambda / d \cos \theta$$

where B_c is the line broadening, K_1 a constant (0·893), λ the wave length, d the crystallite size, and θ the Bragg angle. More refined treatment can also yield a particle size distribution, but the method is insensitive to particles less than about 5 nm because the broadening is then excessive.

Ideally all three methods should be used side by side, and, when this has been done, the agreement between them is often satisfactory. X-ray line broadening and gas chemisorption are complementary in that the first becomes less sensitive, and the latter more sensitive, as particle sizes decrease.

There has been much speculation whether specific activity (that is, the activity per unit area) of metal particles is constant down to the smallest particle sizes possible. The answer seems to be that it depends on the reaction used. For many reactions it is seemingly constant, and these reactions are called *facile*, or *structure-insensitive*. A few reactions (especially hydrogenolysis of carbon–carbon bonds) show a specific activity which increases with decreasing particle size; these are called *demanding*, or *structure-sensitive*, reactions.

4.7. Unsupported metals for use in liquid media

Although, as noted in the last section, supported metals are generally more convenient to use than unsupported counterparts, there is nevertheless some interest in and some applications for finely divided unsupported metals. Colloidal platinum is for example readily prepared by reducing a dilute aqueous solution of chloroplatinic acid with sodium citrate, and the resultant metal particles are of almost uniform size with an average size of about 1·5 nm. The ability to prepare such small and uniform particles is attractive for fundamental research. A very fine dispersion of platinum in non-aqueous solvents can also be made by *in situ* reduction with hydrogen of Adams platinum oxide, $PtO_2 \cdot H_2O$, made by heating a platinum salt in molten sodium nitrate.

A further useful form of unsupported metal is *Raney nickel*, sometimes referred to as *skeletal nickel*. This is prepared by forming an alloy of nickel and aluminium having equal parts by weight, from which the aluminium is dissolved, immediately before use, with alkali. The resulting nickel, containing traces of residual aluminium and alumina, has a very open structure of high surface area. Raney nickel has found use in fat-hardening (see Chapter 5). Other metals can also be prepared in this form.

QUESTIONS

4.1. For which types of reactor is it (a) important and (b) unimportant to have catalyst particles within a closely controlled size range?

4.2. In the case of a supported metal catalyst in pelleted form, what considerations affect the degree of penetration of metal into the pellet? How would you prepare a nickel on pelleted alumina catalyst having the metal (a) very close to the surface of the pellet, and (b) uniformly distributed throughout it?

4.3. In a fluidized-bed reactor, how do you expect the particle size to affect the flow rate of gas at which fluidization occurs?

4.4. In the case of a reactor where it is necessary to isolate an intermediate product in high yield, would it be preferable to use a catalyst or support of high porosity or of low porosity? If the latter, should the particles be large or small?

4.5. What considerations affect the maximum height that is possible for a fixed-bed reactor?

5. Catalysis in the food industry

Hydrogenation of carbon–carbon multiple bonds, and the production of margarine

5.1. Hydrogenation and isomerization of olefins

THE point has already been stressed (section 3.2) that measurement of orders of reaction is by no means sufficient to define a reaction mechanism, since many alternative combinations of unit steps can lead to the same overall dependence of rate on pressures of reactants. Indeed there seems to be no generally accepted statement of what constitutes a reaction mechanism, and because of the various levels at which understanding may be sought it is important to have a working definition. The following is therefore proposed. 'The mechanism of a catalysed reaction is adequately understood if the following points are established beyond reasonable doubt: (i) the nature of all the species participating in the reaction; (ii) the qualitative modes of their interaction contributing significantly to the total reaction; (iii) quantitative aspects of these interactions expressed on a relative but not absolute basis; and (iv) the formulation of the rate-determining step.'

We have already discussed the chemisorption of hydrogen and of olefins (Chapter 2), and so we may begin our discussion of the mechanism of olefin hydrogenation by describing the chemisorbed states of the reactants as follows.

```
  H    H              RHC—CHR
  |    |               |    |
—M—M—               —M—M— .
  |    |               |    |
hydrogen atoms          olefin
```

This symbolism is unwieldy and it is preferable to follow the widely adopted convention of using an asterisk to represent the adsorption site, so that we have

```
      H                  RHC—CHR.
      *                   *   *
chemisorbed H atom    chemisorbed olefin
```

The important question which now requires attention is whether the two hydrogen atoms are added to the olefin simultaneously or consecutively, and if the latter what intermediate species results. The answer to this question is provided by results obtained in studying the hydrogenation of 1-butene in a static system over palladium/alumina at 310 K. Figure 5.1 shows how the

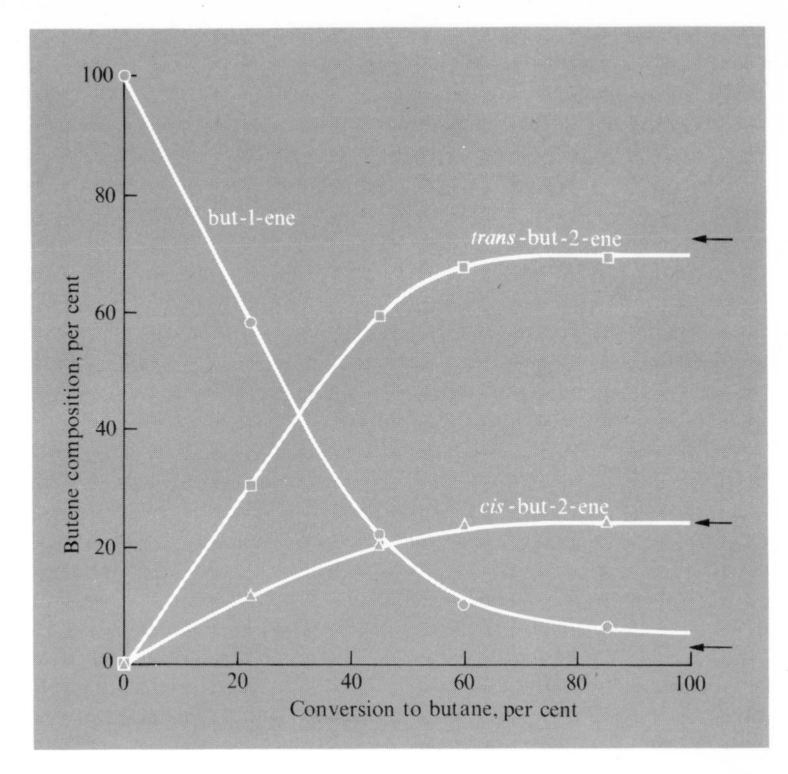

FIG. 5.1. Isomerization of 1-butene over Pd/Al_2O_3 in a static system at 310 K. (Arrows indicate equilibrium proportions expected from thermodynamic calculation.)

composition of the olefin changes as the hydrogenation proceeds; the two isomeric 2-butenes (the *cis*- and *trans*-isomers) are formed by a process called *double-bond migration.* Indeed whichever isomer one starts with, the other two are formed from it either by double-bond migration or by *cis–trans isomerization,* both of which processes are referred to as *olefin isomerization.* Where the isomerization occurs sufficiently quickly, the olefins can achieve their thermodynamic equilibrium before they are completely hydrogenated, as indeed is the case in Fig. 5.1.

We must now try to formulate a mechanism to account for the formation of isomerized olefins. If one of the two hydrogen atoms reacts with adsorbed 1-butene before the other, we see from Fig. 5.2 that there are two possible species which can be formed: if the first hydrogen atom adds to the terminal carbon atom we get a 2-butyl radical, but if it adds to the second carbon atom a 1-butyl radical is formed. We then postulate that this hydrogen atom

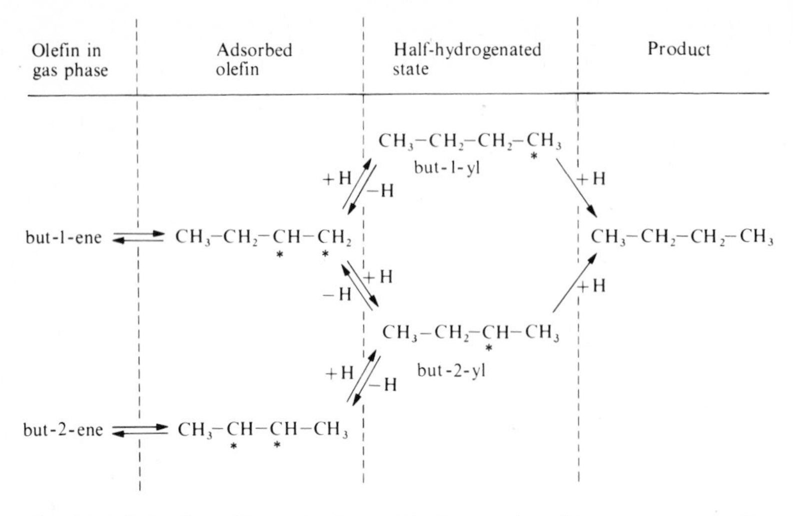

FIG. 5.2. Mechanism of isomerization and hydrogenation of butenes on a metallic catalyst.

addition is a reversible process. However, in the case of the 1-butyl radical, loss of a hydrogen atom can only re-form adsorbed 1-butene, whereas with the 2-butyl radical it can occur from either the first carbon atom (giving adsorbed 1-butene again) or from the third carbon atom, in which case adsorbed 2-butene results.

The interconversion between adsorbed *cis*- and *trans*-2-butene can be explained in a similar way. How this can occur is most clearly shown by handling molecular models, but an attempt to describe it is made in Fig. 5.3. If a hydrogen atom (H_B) is added to adsorbed *cis*-2-butene (and whichever end of the double bond it adds to, only a 2-butyl radical can be formed), and if the same hydrogen atom is subsequently abstracted, then adsorbed *cis*-2-butene is re-formed; but if it is the other hydrogen atom (H_A) which is removed, then adsorbed *trans*-2-butene results. In favourable circumstances, equilibrium proportions of all three isomers may be obtained irrespective of which isomer is taken as the reactant. Figure 5.2 indicates how 1-butene may arise from *cis*- or *trans*-2-butene. It must be remembered that the isomerized product will only be observed if it desorbs from the surface into the gas phase. If the olefin is so strongly adsorbed that desorption is rare, then isomerization will not be seen even though the reaction has occurred on the surface.

The alkyl radical intermediates are frequently referred to as *half-hydrogenated states*. This is because addition of the second hydrogen atom to them breaks the carbon–metal bond and gives the saturated hydrocarbon in the gas phase (see Figs. 5.2 and 5.3).

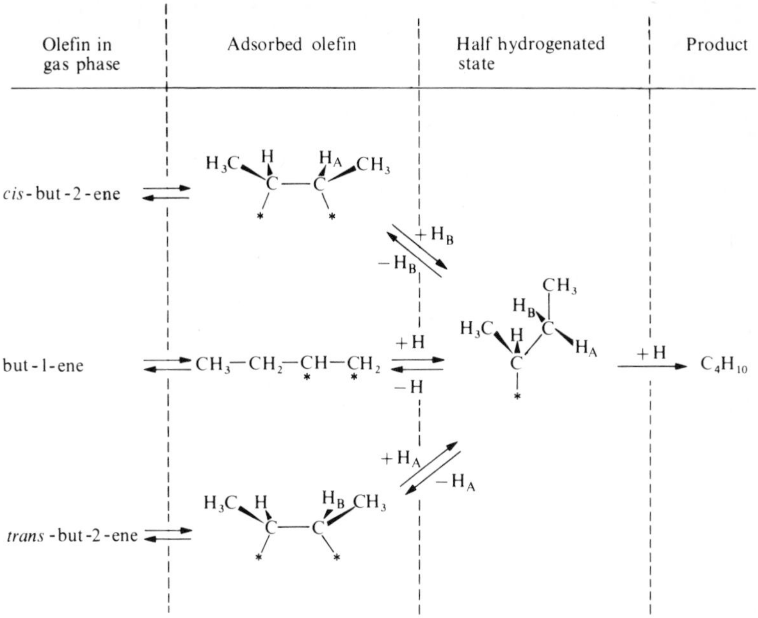

Olefin in gas phase	Adsorbed olefin	Half hydrogenated state	Product
cis-but-2-ene ⇌	H_3C, H, H_A, CH_3 on C—C (adsorbed, * *)		
	$+H_B$ / $-H_B$		
but-1-ene ⇌	$CH_3-CH_2-CH-CH_2$ (* *) $\underset{-H}{\overset{+H}{\rightleftharpoons}}$	CH_3, H_B, H_3C, H, C, H_A (C *)	$\xrightarrow{+H}$ C_4H_{10}
	$+H_A$ / $-H_A$		
trans-but-2-ene ⇌	H_3C, H, H_B, CH_3 on C—C (adsorbed, * *)		

FIG. 5.3. Mechanism of *cis–trans* isomerization of olefins on a metallic catalyst.

Analogous processes occur with ethylene, but these require the use of deuterium as a stable isotopic tracer to reveal them. It is observed that when ethylene reacts with deuterium over a metallic catalyst there are formed (i) deuterated ethylenes, with one to four deuterium atoms, (ii) deuterated ethanes with zero to six deuterium atoms, and (iii) hydrogen and hydrogen deuteride (HD). The experimental procedure by which the products are analysed is complicated, but essentially they are separated into the chemical components by preparative gas–solid chromatography, and each component is then examined by mass spectrometry. The quantities of deuterated ethylenes and of hydrogen and HD which are formed, and the breadth of the distribution of deuterated ethanes varies considerably with temperature, reactant pressures, and especially with the nature of the catalyst (see below). Some typical results are shown in Table 5.1. The mechanism that can account for the formation of all the observed products, both qualitatively and quantitatively, involves the reversible formation of an ethyl radical as the half-hydrogenated state (Fig. 5.4). Reiteration of the basic steps can give complete exchange of the hydrogen atoms in ethylene for deuterium atoms, while interaction of variously deuterated ethyl radicals with either hydrogen or deuterium atoms accounts for all the deuterated ethanes, including C_2H_6. Hydrogen and hydrogen deuteride

TABLE 5.1

Percentage distribution of products from the reaction of ethylene with deuterium over some Group VIII metals on γ-alumina

		Ethylenes				Ethanes							
Metal	Temp., K	C_2H_3D	$C_2H_2D_2$	C_2HD_3	C_2D_4	C_2H_6	C_2H_5D	$C_2H_4D_2$	$C_2H_3D_3$	$C_2H_2D_4$	C_2HD_5	C_2D_6	M*
Pd	310	41·5	8·9	1·6	0·1	24·5	16·1	5·6	1·3	0·3	0·1	0·0	0·69
Pt	327	9·2	1·2	0·2	0·0	19·3	28·2	25·6	8·7	4·9	2·3	0·4	1·55
Ir	360	8·1	2·8	0·7	0·2	4·6	25·7	30·0	12·5	9·4	4·8	1·2	2·18
Rh	350	61·7	8·7	2·1	0·1	3·5	7·7	15·1	0·9	0·2	0·0	0·0	1·50

* Average number of D atoms in the ethanes as defined by $C_2H_{6-M}D_M$.

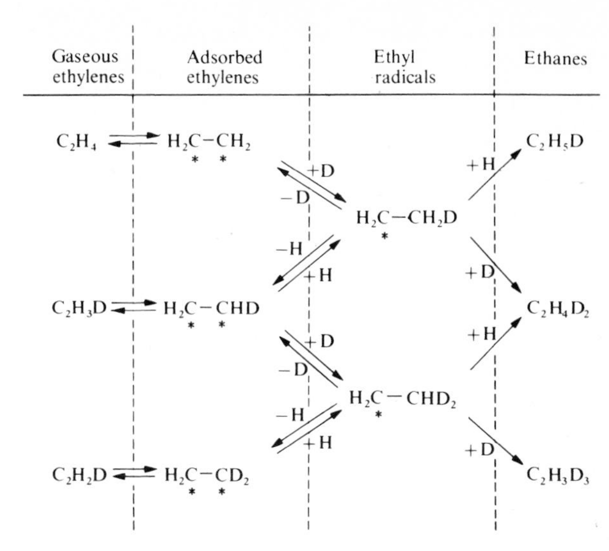

FIG. 5.4. Mechanism of the reaction of ethylene with deuterium over a metallic catalyst.

are formed by recombination and desorption of hydrogen and deuterium atoms.

The rate of formation of isomerized olefins (r_i) relative to the rate of hydrogenation (r_h) varies greatly from one metal to another. The ratio r_i/r_h is high over nickel, palladium, ruthenium, and osmium, but much lower over platinum and iridium: this is partly at least because ethylene is more strongly adsorbed on the last two metals than on the former group. The same effect is shown with the rate of formation of deuterated ethylenes compared with the rate of formation of ethane, although detailed kinetics are somewhat different.

It is clear that the wealth of mechanistic information revealed by studies of isomerization and use of a stable isotope could not be seen merely by measuring orders of reaction. Although exchange and isomerization are not *necessary* precursors to hydrogenation, many molecules will have experienced these reactions before suffering final hydrogenation, and they are therefore a legitimate part of the reaction mechanism.

We must note in passing that certain oxides (e.g. ZnO and Cr_2O_3) are tolerably good hydrogenation catalysts, and that the reaction mechanism is considerably simpler than on metals. For example, reaction of ethylene with deuterium gives only the expected simple addition product 1,2-dideuteroethane.

5.2. Hydrogenation of multiply unsaturated hydrocarbons

The hydrogenation of molecules containing multiple unsaturation (e.g. acetylene and butadiene) exhibits features not shown in the reactions of mono-olefins. The principal difference is that the reaction can over certain catalysts occur in two quite well separated stages, viz.

$$C_2H_2 \rightarrow C_2H_4 \rightarrow C_2H_6$$

or

$$C_4H_6 \rightarrow C_4H_8 \rightarrow C_4H_{10}.$$

The degree to which the two stages are separated is expressed by the *selectivity* of the reaction, defined as (for acetylene)

$$\text{selectivity} = \frac{\text{rate of formation of } C_2H_4}{\text{rate of formation of } C_2H_4 + C_2H_6}$$

and analogously for butadiene. A factor aiding the separation of the two stages is that acetylene and butadiene are much more strongly chemisorbed on metals than the mono-olefins to which they give rise, and so the latter once formed and desorbed from the surface cannot gain readmittance to the surface until almost all the parent compound has reacted.

By analogy with what has been written in the last section on reaction mechanisms, the outline of the mechanism of the hydrogenation of acetylene is as shown in Fig. 5.5: the half-hydrogenated state is the vinyl radical. As with

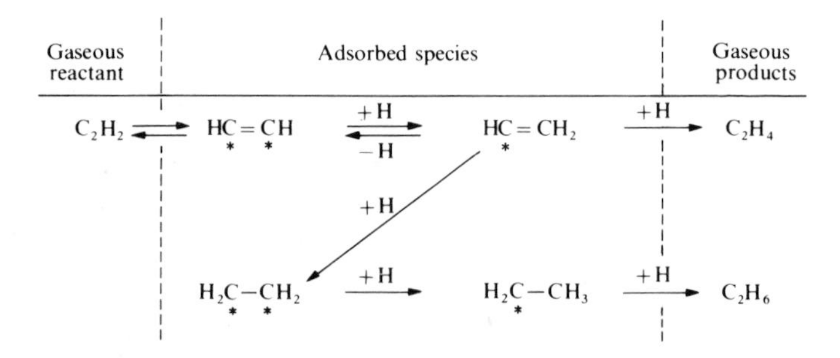

FIG. 5.5. Mechanism of hydrogenation of acetylene over a metallic catalyst.

olefin isomerization and exchange, selectivity depends greatly on the catalyst and the conditions employed, usually decreasing with increase of hydrogen pressure and decrease of temperature. Table 5.2 shows values of selectivities for butadiene hydrogenation over various alumina-supported catalysts. It is interesting to note that iridium and platinum, which were poor for olefin

TABLE 5.2

Selectivities in the formation of butenes from butadiene over various metals on γ-alumina

Metal	Selectivity	Temperature/K
Fe	0·98	470
Ru	0·84	320
Os	0·58	320
Co	1·00	350
Rh	0·84	320
Ir	0·35	320
Ni	1·00	350
Pd	1·00	320
Pt	0·63	320
Cu	1·00	370

isomerization, show low selectivities, while other metals which gave faster olefin isomerization tend to have quite high selectivities. This is because the rate of olefin desorption is an important factor in achieving high selectivity and fast isomerization or exchange rates.

Palladium is unique in frequently exhibiting almost complete selectivity over a wide range of conditions, as well as very high activity, in acetylene hydrogenation. For this reason it is used industrially to remove by hydrogenation traces of acetylenes and diolefins from gas streams containing mono-olefins, where the presence of these traces would be detrimental to subsequent processing.

Double-bond migration does not occur with acetylenes or conjugated diolefins, but non-conjugated olefins readily isomerize under hydrogenation conditions to their conjugated counterparts, e.g.

$$\underset{\text{1,4-pentadiene}}{H_2C{=}CH{-}CH_2{-}CH{=}CH_2} \rightarrow \underset{\text{1,3-pentadiene}}{H_2C{=}CH{-}CH{=}CH{-}CH_3}.$$

5.3. The objectives of fat hardening

The foregoing discussion of the mechanism of hydrogenation of unsaturated hydrocarbons serves as an introduction to the industrially important subject of fat hardening, the process whereby animal and vegetable oils are converted into edible fats.

Nature provides us with a great variety of such oils which have considerable nutritional value. Possible applications for these materials, after such treatment as may be appropriate, are in domestic cooking, in large-scale food manufacture (e.g. cakes and biscuits), as components in animal feedstuffs, and as butter substitutes (e.g. margarine). These animal and vegetable oils are complex mixtures of tri-esters of glycerol which may be represented as

$$\begin{array}{l} CH_2O{\cdot}CO{\cdot}R^1 \\ | \\ CHO{\cdot}CO{\cdot}R^2 \\ | \\ CH_2O{\cdot}CO{\cdot}R^3 \end{array}$$

The alkyl groups R^1, R^2, and R^3 are unbranched, and contain 15, 17, or occasionally some higher or lower odd number of carbon atoms. It is usual to describe the chemical composition of any oil in terms of the *fatty acids* that result on hydrolysis. We shall confine our attention chiefly to the C_{18} fatty acids ($R{\cdot}CO_2H$ where R contains 17 carbon atoms): these usually contain one or more double bonds disposed as shown in Table 5.3. For our purpose the

TABLE 5.3

Structural characteristics of some C_{18} *acids*

Linolenic acid	$CH_3{\cdot}CH_2$ $\quad CH_2 \quad CH_2 \quad CH_2[{\cdot}CH_2]_6{\cdot}CO_2H$ $CH{=}CH \quad CH{=}CH \quad CH{=}CH$
Linoleic acid	$CH_3[{\cdot}CH_2]_3{\cdot}CH_2 \quad CH_2 \quad CH_2[{\cdot}CH_2]_6{\cdot}CO_2H$ $CH{=}CH \quad CH{=}CH$
Oleic acid	$CH_3[{\cdot}CH_2]_6{\cdot}CH_2 \quad CH_2[{\cdot}CH_2]_6{\cdot}CO_2H$ $CH{=}CH$
Elaidic acid	$CH \quad CH_2[{\cdot}CH_2]_6{\cdot}CO_2H$ $CH_3[{\cdot}CH_2]_6{\cdot}CH_2 \quad CH$
Stearic acid	$CH_3[{\cdot}CH_2]_{16}{\cdot}CO_2H$

manner in which the fatty acid moities ($R{\cdot}CO{\cdot}$) are distributed between the available positions on the glycerol molecule is irrelevant. We note in passing that the terminology of the industry has been to call the trilinolenyl ester of glycerol 'trilinolenin'. Thus we may also have dilinoleolinolenin, oleodistearin etc. etc. The analysis for these individual glycerides was, until the advent of gas–liquid chromatography, a most complex procedure.

The two principal sources of these useful oils are (i) the seeds of plants, which provide the vegetable oils such as linseed oil, olive oil, castor oil, palm oil, and soybean oil, and (ii) the tissues of animals, especially fish, which provide for example whale oil and cod liver oil. The production of soybean oil much exceeds that of any other vegetable oil, and is about 5 million tons per annum. The soybean contains about 20 per cent oil which is extracted by pressing, the residue being used for animal feedstuffs and for meat substitutes. The oil has the following approximate composition: linolenic acid, 8%; linoleic acid, 50%; oleic acid, 27%; stearic acid, 4%; palmitic acid, 10%. Palmitic acid is a C_{16} fully saturated acid. One of the principal problems encountered in treating animal and vegetable oils is the considerable seasonal variation that occurs in the average number of double bonds per molecule in a given oil. Different oils vary widely from each other, linseed oil being especially rich in polyunsaturated species.

Unfortunately the oils as provided by nature are not immediately suited to many of their potential applications. The two important reasons for this are their unsavoury flavour and their inappropriate physical form. The chief cause of the former is that on storage in air the polyunsaturated acids suffer oxidation especially at the activated methylene group between double bonds. The products have an evil taste and the material is said to have gone *rancid*. The occurrence of this process means that these oils have very short shelf lives, which is a distinct disadvantage to their commercial use. While there is a large demand for liquid oils (e.g. for cooking, and salad oils), a cheap substitute for butter is also desirable, and this depends upon finding a means of producing something of similar consistency, with if possible an even wider range of temperature in which 'spreadability' is good. Fortunately both of these disadvantages can be quite fully overcome by the technique of partial catalytic hydrogenation.

The purpose of the hydrogenation treatment is to diminish as far as possible the concentration of linolenic acid, to decrease somewhat that of linoleic acid, to increase thereby the concentration of oleic acid while not forming too much stearic acid or elaidic acid. The desirable properties of margarine include a long shelf life, but complete elimination of linoleic acid is undesirable because of the part it is thought to play in preventing diseases of the blood circulation system. Some formation of the higher melting stearic or elaidic acids is essential to secure satisfactory consistency, but stearic acid is not readily digested. The desired product is therefore a compromise determined by a delicate balance of the above factors. In chemical terms the wish is to obtain a controlled partial hydrogenation of the multiple carbon–carbon bonds, without excessive *cis–trans* isomerization of oleic to elaidic acid, and without much complete reduction to stearic acid. The practical means of achieving this end, with some comments on mechanisms, are considered in the following section.

5.4. Theory and practice of hydrogenation of edible oils

The objectives outlined above can be at least partly achieved by catalytic hydrogenation: we need to recall some of the ideas introduced in sections 5.1 and 5.2 to be able to understand how. Success depends on two factors: the first is the ability of non-conjugated species such as linolenic acid and linoleic acid to isomerize to their conjugated counterparts, e.g.

—C=C—C—C=C— → —C—C=C—C=C—

linoleic acid → isolinoleic acid

—C=C—C—C=C—C—C=C— → —C=C—C—C=C—C=C—C—

linolenic acid → partly conjugated linolenic acid

↓

—C—C=C—C=C—C=C—C—

fully conjugated linolenic acid

The second important factor is that conjugated dienes and especially trienes are more strongly adsorbed on the catalyst surface than monoenes (e.g. oleic acid), and hence hydrogenated preferentially. To obtain the most selective reduction on linolenic acid it is important to use the right metal: for a great many years, only nickel has been employed industrially, and reference to Table 5.2 shows that its selectivity should be good, as indeed it is. The progress of a typical nickel-catalysed hydrogenation of an oil such as soybean oil is sketched in Fig. 5.6. The point at about which hydrogenation would be stopped is shown by the broken line. The conjugated isomers of linolenic acid and linoleic acids are not shown; their concentrations often remain very low, because they are hydrogenated more quickly than they are formed.

It is perhaps unfortunate that nickel, which is a very acceptable catalyst on grounds of price and selectivity, is also very good at *cis–trans* isomerization. The appearance of *trans*-isomers of oleic and iso-oleic acids is shown in Fig. 5.6: as noted above, however, they do help to harden the product, which effect can otherwise be achieved only by forming the undesirable stearic acid.

Some attention has recently been given to the use of copper as a catalyst for fat-hardening. Copper has the advantage that it is inactive for reduction of mono-olefins, but, regrettably, small traces of copper remaining in the product catalyse the autoxidation of the unsaturated compounds, leading to poor flavour stability.

About 90 per cent of all industrial operations in the edible-oil hydrogenation field are conducted as batch operations. There are a few plants using trickle-column reactors (see section 4.3), but the seasonal variation in oil composition referred to above is a nuisance in a continuous process, and further the large glyceride molecules do not readily diffuse inside the pores of pellets or granules, so that only very poor selectivity is achieved.

Batch reactors are designed to give good contact between the hydrogen gas, the oil (in which hydrogen is only very slightly soluble), and the catalyst which

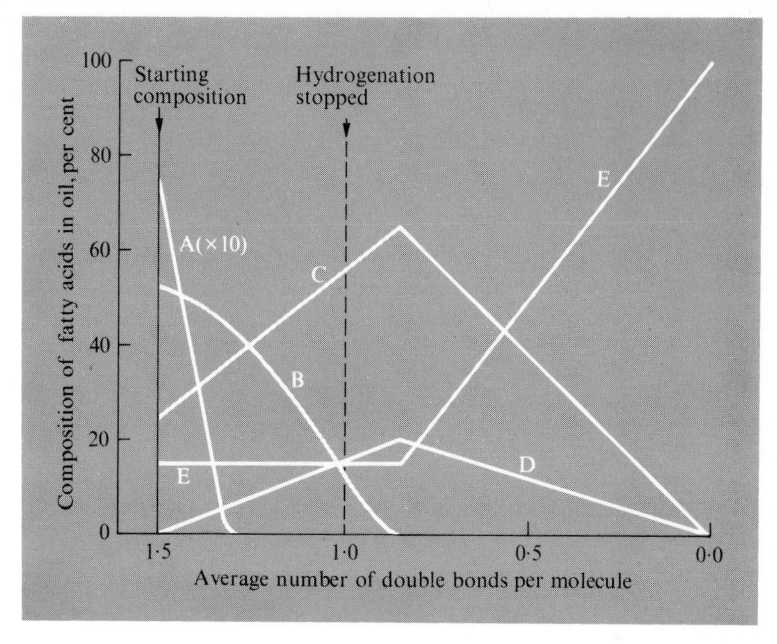

FIG. 5.6. Course of hydrogenation of soybean oil with a nickel catalyst (schematic). A: linolenic acid (scale ×10); B: linoleic acid; C: oleic acid and iso-oleic acids; D: elaidic acid and other trans-isomers; E: stearic acid.

is suspended in the oil. The reactor is typically cylindrical, with a central stirrer axis bearing two or three impellers. Such a reactor may be operated on a 'dead end' basis, in which the exit valve is closed during hydrogenation, the gas being introduced by a sparger at the bottom. Continuity of reaction relies on good dispersion of small hydrogen bubbles in the oil, this being achieved by the action of stirrers. The 'dead end' mode of operation is mechanically simple, but is sensitive to design of the stirrers and to baffling arrangements, and also to the depth of liquid in the reactor.

A batch reactor can also be operated in a 'gas recirculation' mode, in which a loop containing a compressor returns gas from the top of the reactor (the 'head space') to the bottom. It can be purified if necessary in this return loop. This arrangement is mechanically more complex than the 'dead end' mode, but does not suffer from the latter's disadvantages noted above. The reaction is highly exothermic, and a well-designed reactor is equipped with a considerable array of cooling coils.

After the desired degree of hydrogenation has been obtained, the hydrogen supply is stopped; the charge is cooled from reaction temperature (usually around 450 K) to below 370 K and is then filtered. Owing to the very small size of the catalyst particles, hydrocyclones and centrifuges have been found inadequate to separate the product from catalyst, and old-fashioned plate and frame filter presses have to be used.

Improved selectivity (and increased yield of *trans*-isomers) results from operation in a regime where mass-transport limitation of hydrogen exists. Since much of the reaction occurs within the pores of the small catalyst particles, provided they are sufficiently wide, it will naturally be diffusion limited with respect to hydrogen, so no particular precautions need be adopted other than proper catalyst design.

A variety of forms of nickel are used. Nickel/silica is widely employed, and the metal may also be introduced as nickel formate dispersed in hardened fat. The formate decomposes to metallic nickel in the reactor, thus avoiding contamination of the catalyst before it comes into use.

QUESTIONS

5.1. A number of technical terms were used in describing fat hydrogenating in a batch reactor, but their meanings were not explained, e.g. impeller, sparger, hydrocyclone, centrifuge, and plate and frame filter press. If you do not understand these terms, consult a chemical engineering text-book to find out.

5.2. Why are oxides like ZnO and Cr_2O_3 not used as catalysts for fat hardening? Why is nickel used in preference to (a) cobalt and (b) palladium?

5.3. The reaction of ethylene with hydrogen typically obeys the equation

$$r = kP_{H_2}^{1}P_{C_2H_4}^{0}$$

Assuming the existence of ethyl radicals as intermediates, and using the method of stationary states, see how many 'mechanisms' are consistent with this equation.

5.4. The selectivity observed in acetylene hydrogenation decreases with decreasing temperature and increasing hydrogen pressure. How do these observations fit with the outline mechanism presented in this Chapter?

5.5. What products would you expect to be formed during hydrogenation of the following molecules?

(a) 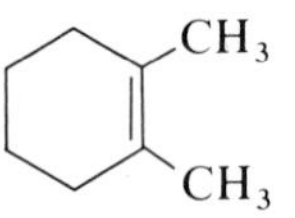 CH_3 CH_3 (b) 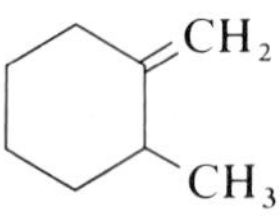 CH_2 CH_3 (c) 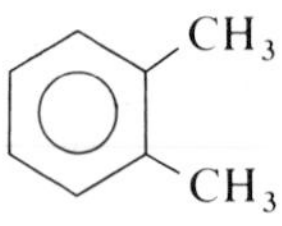 CH_3 CH_3

6. Catalysis in the fine-chemicals industry

6.1. Introduction

It is not easy to define what constitutes the *fine-chemicals industry*. This sector of industry is concerned with manufacturing, processing, and transforming a great variety of organic compounds which find use as pharmaceutical and medicinal products, as dyestuffs, food additives, cosmetics, and in photography and other methods of reproduction, to name but a few. (See E.S. Stern (ed) *The chemist in industry* 1 (OCS 12) and 2 (OCS 17).) The products of the pharmaceutical industry are extremely diverse, ranging from simple analgesics and antipyretics such as aspirin and phenacetin, through sulphonamides to penicillins, aureomycin, and the tetracyclines. The compounds involved are often structurally complex, of high molecular weight, involatile, and sensitive to heat. The aid of catalysis is frequently invoked to secure desired transformations under comparatively mild conditions. So great is the range of chemicals employed that we can only here discuss some general principles, and a few typical reactions which find application in a number of situations.

For the reasons just mentioned, most of the reactions of interest to the fine-chemicals industry necessitate the use of the reactant dissolved in a suitable solvent, or more rarely of the pure liquid reactant. Thus finely divided catalysts which can be easily kept in suspension must be used, or occasionally trickle-column reactors (see section 4.3). Temperatures must usually be kept as low as possible, but rates of reactions involving a gaseous reactant (hydrogen or oxygen) may be improved by using superatmospheric pressure (see section 3.4). Diffusion limitation of rates (section 3.4), which is discussed further below, is often experienced.

We now consider some examples of reactions frequently encountered in fine-chemicals operations.

6.2. Catalytic hydrogenation and related reactions

The process of adding hydrogen across a double bond is one of the most frequently encountered catalytic reactions in the fine-chemicals industry. The organic molecules of interest usually have several reactive groups, and hydrogenation can therefore conceivably give several different products of which perhaps only one is wanted. Careful choice of all the variables of the system (catalyst, solvent, temperature, pressure, agitation etc.) is therefore required to maximize the yield of this desired product.

Some unsaturated groups differ widely in reactivity, and little difficulty is encountered in finding conditions where their reactions can be clearly separated. For example, the benzene ring is not easily reduced, and the selective reduction of styrene to ethylbenzene presents no problems. Many other groups substituted onto an aromatic ring are also selectively reduced with ease: the conversions of nitrobenzene or nitrosobenzene to aniline, of benzaldehyde to benzyl alcohol and of benzonitrile to benzylamine are readily effected under appropriate conditions.

Correct choice of catalyst is most important. Although Raney nickel is often used, the need to employ the mildest possible conditions, and hence the most active catalyst, often makes the choice fall on one of the noble Group VIII metals, of which palladium is the least expensive, one of the most active, and the most widely used. Platinum is generally less selective in its behaviour than palladium. Rhodium and ruthenium find use, especially for laboratory-scale work, in reducing aromatic rings, which they can do at room temperature and elevated pressure. Selectively poisoned and sulphided catalysts are finding increased use. Although there are some general guidelines, one or two of which are hinted at above, the final choice of catalyst is determined by experience of its selectivity, its resistance to poisoning, and other practical considerations that are not readily predicted.

Choice of solvent is made on an even more empirical basis. Neutral solvents such as methanol, ethanol, hexane, and cyclohexane are generally useful, but the solubility of hydrogen in hydrocarbons is lower than in alcohols, and hence reactions conducted in hydrocarbons more readily become diffusion limited. Acidic solvents (e.g. acetic acid) are beneficial when nitrogen bases, which may strongly adsorb on the catalyst, are reaction products; basic solvents are useful when products are acidic. Solvents which can themselves be hydrogenated (e.g. acetone) are not often used, and those having a high vapour pressure at the reaction temperature are not suitable because the maximum partial pressure of hydrogen is thereby limited.

We return now to the important question of selectivity. One of the forms in which it is most often encountered is when an intermediate product, or another functional group in the molecule, is destroyed by hydrogenolysis. Thus while it was remarked above that hydrogenation of benzaldehyde to benzyl alcohol was easy, we must now note that the hydrogenolysis of benzyl alcohol to toluene is also quite facile, although it can be suppressed by the addition of small quantities of a nitrogen base (e.g. quinoline) which adsorbs competitively. Hydroxyl and alkoxy groups, and halogen atoms, on aromatic rings are also readily cleaved by hydrogenolysis, and thus for example the conversion of a chloronitrobenzene to a chloroaniline is no simple problem. Unfortunately there is no substitute for experience in this jungle.

Some reactions are notoriously difficult to effect. The selective reduction of conjugated aldehydes to the corresponding alcohol, e.g. of cinnamaldehyde to cinnamyl alcohol, is one such; reduction of the olefinic bond normally proceeds preferentially (see Fig. 6.1). The carboxyl group ($—CO_2H$) is extremely difficult to reduce under mild conditions, although dirhenium

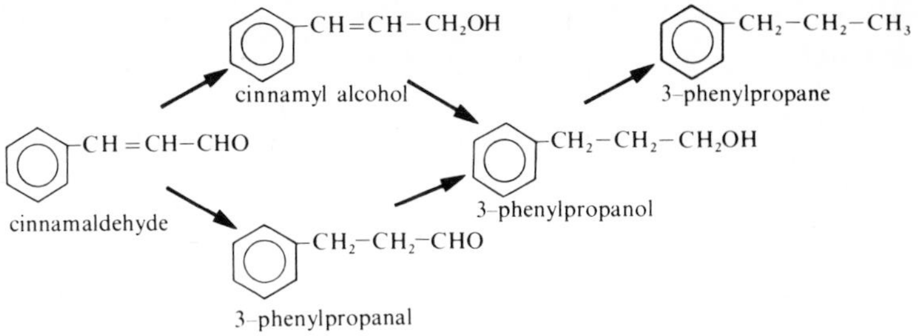

FIG. 6.1. Pathways in the reduction of cinnamaldehyde.

heptasulphide (Re_2S_7) is effective under forcing conditions of high temperature and pressure.

Surprising reactions sometimes occur during catalytic hydrogenation. The reduction of nitrobenzene over platinum in dilute sulphuric acid gives a good yield of *p*-aminophenol, and in anhydrous hydrogen fluoride gives *p*-fluoroaniline. Another interesting and useful reaction is reductive alkylation: primary or secondary amines or ammonia react with aldehydes or ketones in the presence of hydrogen and a suitable catalyst (e.g. palladium) to give a new amine. Thus for example the reaction of 2-methylcyclohexanone with ammonia gives 2-methylcyclohexylamine, and of cyclohexanone with ammonia gives dicyclohexylamine. The reasons for this behaviour are not clearly understood.

Hydrogen transfer agents are sometimes used in place of hydrogen for small-scale work. Thus, in the presence of a catalyst, hydrogen atoms from isopropanol or cyclohexane can transfer to a hydrogen acceptor with the formation of acetone or benzene respectively. The catalytic disproportionation of cyclohexene to cyclohexane and benzene is an example of an intermolecular hydrogen transfer.

6.3. Diffusion limitation in three-phase systems

The basic principles of mass-transport limitation, or diffusion limitation, were outlined in section 3.4; in view of its frequent or often unsuspected occurrence in catalytic hydrogenation where a liquid phase is present, some further consideration is essential.

Figure 6.2 shows in a formal way how the rate of hydrogen consumption varies with agitation efficiency (expressed as rpm for a stirred reactor or as vibrations per minute for a small shaken reactor) with different weights of catalyst. It is often found that a minimum rate of stirring or shaking has to be exceeded before the liquid surface begins to break up and thus before reasonably fast rates of reaction are observed. There is therefore an intercept of about 500 rpm on the stirring rate axis, the precise value depending on the presence of baffles and stirrer design. Qualitatively, the greater the weight of catalyst

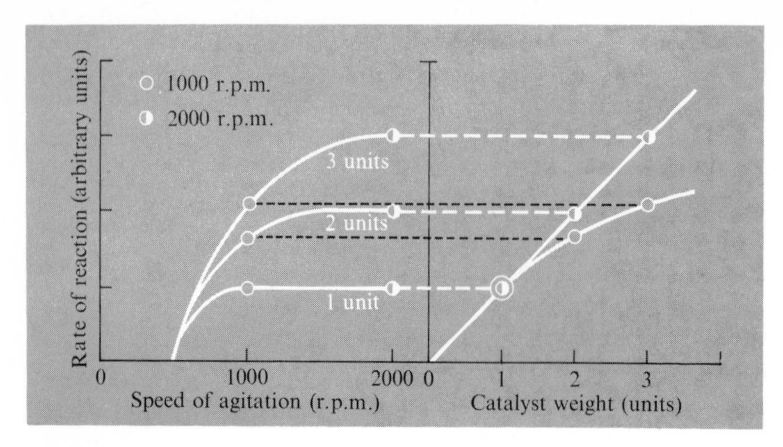

FIG. 6.2. Dependence of reaction rate on speed of agitation using different quantities of catalyst (left) and on catalyst weight (right).

used, or the higher the concentration of active component in it, the faster must the reactor be shaken or stirred to overcome diffusion limitation. This plot may then be easily translated into one of reaction rate versus catalyst weight (see also Fig. 6.2), from which the non-linear dependence of rate on catalyst weight under conditions of diffusion limitation is readily apparent.

Other symptoms of diffusion limitation, and methods in principle of overcoming the problem, were outlined in section 3.4. The diffusion of organic molecules in solution is rarely rate-limiting, but owing to the low solubility of hydrogen in most liquids it is almost always transport of hydrogen to the catalyst that holds the reaction up. The use of (a) solvents with high solubilities for hydrogen, (b) low temperatures, and (c) high pressures maximizes the concentration of dissolved hydrogen and minimizes the chance of diffusion becoming the slowest step of the total sequence. Low catalyst weights, or low concentrations of active components, and good agitation are the other recommended conditions for avoiding diffusion limitation.

7. Catalysis in the petroleum industry

The cracking of mineral oil and the reforming of petroleum

7.1. The objectives of the industry

THE earth provides abundant supplies of hydrocarbons, either as 'natural gas' or as crude oil. The former consists of low-boiling hydrocarbons (almost pure methane in the case of 'North Sea gas', and 90 per cent methane with 10 per cent ethane, propane, and butane in the case of North American natural gas) and is predominantly used directly as a fuel for domestic and industrial purposes. Crude oil contains hydrocarbons covering a great range of molecular weight and structural complexity, the composition varying considerably with the source. Most crude oils also contain sulphur and nitrogen as heterocyclic compounds, with traces of metals (especially vanadium), the quantities again varying greatly with the origin of the oil. For our purposes we may regard crude oil as containing principally linear and alicyclic hydrocarbons of various molecular weights.

The following is a very brief summary of the operations that ensue. Crude oil is first separated into a fraction which is volatile at about 670 K and a non-volatile 'residual oil'. The volatile fraction, typically three-quarters of the whole, is fractionally distilled, giving the following cuts in order of decreasing volatility: (1) C_1–C_4 hydrocarbons, (2) light gasoline, (3) heavy gasoline or naphtha, (4) kerosene, and (5) light gas oil. The 'residual oil' is distilled under reduced pressure to give heavy gas oils and a further residue of high-boiling material which may be suitable for use as a fuel oil. Products of this distillation may be subjected to catalytic cracking (see below) or may be used for production of lubricating oils.

Annual world production of crude oil exceeded 2000 million tons in 1968. By far the greater part is used as fuels and only a small fraction as raw materials for manufacture of chemicals: in Europe the figures are about 90 per cent and 3 per cent respectively.

We may classify the applications of petroleum products in the following way: (1) as fuels for aircraft, cars, and diesel engines; (2) as fuels for domestic and industrial heating; (3) as raw materials for the manufacture of chemicals (see Chapter 8); (4) as raw materials for the manufacture of hydrogen and 'town gas' (see Chapter 9); and (5) as lubricating oils, hydraulic fluids, and greases. The technical objectives of the petroleum industry are therefore to separate the components of crude oil and to reform its distillate, to produce a range of products meeting the high specifications demanded in the foregoing applications. In this Chapter we concentrate on the catalytic cracking of heavy oils and on catalytic reforming to give fuels of high octane number suitable for use in cars and aircraft.

7.2. Catalytic cracking

The chief purpose of catalytic cracking is to break high molecular weight hydrocarbons into smaller fragments of suitable volatility to permit their use as fuels. There is however another important criterion for the usefulness of a hydrocarbon fuel in an internal combustion engine, and this is the way it burns in the cylinder. Early petrols suffered from the phenomenon of 'pre-ignition', leading to 'knocking': this was (and still is) overcome by the addition of tetra-ethyl lead as an anti-knocking agent. The combustion characteristics of a hydrocarbon mixture are assessed empirically by assigning the product an 'octane rating'. This is defined as the percentage of iso-octane in an n-heptane + iso-octane mixture to which the mixture under examination is equivalent in terms of its combustion behaviour: the octane ratings of n-heptane and iso-octane are taken to be zero and 100 respectively. Aromatic and branched aliphatic hydrocarbons have higher octane values than alicyclic or linear aliphatic hydrocarbons. Thus the chemical composition of the fuel is extremely relevant to its usefulness.

The cracking of mineral oil fractions was first performed non-catalytically, but it was inefficient in that much material of too low a molecular weight was formed at temperatures where cracking proceeds at a significant rate. Furthermore the octane rating of the product boiling in the useful range was only 70 to 80. Thermal cracking was superseded in the 1930s by catalytic cracking, employing naturally occurring acidic clays of the bentonite type as catalysts. After the Second World War these were replaced by synthetic silica–aluminas of superior performance: the octane rating of the product of catalytic cracking is 90 to 95. This is because some of the reactions which convert linear aliphatic hydrocarbons to products of higher octane value also occur during cracking: these reactions include skeletal isomerization of linear to branched-chain hydrocarbons, dehydrocyclization (e.g. n-hexane to cyclohexane) and dehydrogenation.

One of the major problems encountered in practice is the deposition of carbon on the catalyst through an excessive dehydrogenation of hydrocarbon molecules. This naturally leads to a diminution of catalyst activity, but fortunately the carbon is readily burnt off and the catalyst's activity restored. At one time many plants had two reactors working in parallel, one actually performing the cracking and the other undergoing regeneration: their roles were reversed at regular intervals to maintain high efficiency. This problem was greatly simplified by the introduction of the more recently developed zeolite catalysts (see section 2.9). Catalytic cracking is usually performed in a fluidized-bed reactor (see section 4.3).

The mechanism of catalytic cracking is quite well understood: it involves carbonium ions as intermediates. At the reaction temperature, carbonium ions of long-chain hydrocarbons are unstable with respect to smaller molecules, and so the reaction represented as

$$R_1{-}CH_2{-}\overset{+}{C}H{-}CH_2{-}R_2 \rightarrow R_1{-}\overset{+}{C}H_2 + H_2C{=}CH{-}R_2$$

is favourable. Carbonium ions arise through the reaction of traces of olefin formed by dehydrogenation with protons present on the surface of the acidic catalyst (see section 2.9). Skeletal isomerization also occurs by a carbonium-ion mechanism, since branched carbonium ions are stabler than linear ones, so it is not surprising to find this reaction proceeding in parallel with cracking.

7.3. Catalytic reforming

The products obtained by catalytic cracking using aluminosilicate catalysts still do not have a sufficiently high octane rating to satisfy the demands of high compression-ratio engines, and some further processing is therefore necessary. A substantial breakthrough was made in the early 1950s when it was found that the combination of a metallic component with an acidic one was very effective in producing the desired products. Such substances are referred to as *dual-function* (or *bifunctional*) catalysts, and they operate in the following way. On the metallic component, which is usually platinum, the linear aliphatic hydrocarbon undergoes dehydrogenation to olefin plus hydrogen. The olefin then migrates, probably through the gas phase, to an acidic site where it becomes a carbonium ion and undergoes skeletal rearrangement or ring closure or ring enlargement: it can then lose a proton, becoming an olefin again, and migrate back to the metal where it can be rehydrogenated. This apparently complicated sequence of steps occurs with remarkable facility at temperatures which are usually above 570 K, the operating temperature depending on the concentrations and activities of the two components of the catalyst, on the nature of the feedstock and the product quality which is required. The whole process is called *catalytic reforming*.

Dual-function catalysts have several advantages over simple acidic cracking catalysts. To start with, because they are effective at a lower temperature, they are much less liable to suffer deactivation from carbon deposition. Moreover they operate in the presence of hydrogen, which is added to the hydrocarbon feedstock, and this ensures that the metallic function is kept clean and that carbonaceous residues are not formed. There have been many developments and improvements made in the two decades since the introduction of dual-function catalysts. Some of these were directed towards improving the acidic function; thus silica–alumina containing approximately 10 per cent alumina was replaced by alumina whose surface had been rendered acidic by treatment with a fluorine- or chlorine-containing compound. Easier control of the level of activity could be obtained in this way. More recently it was discovered that the inclusion of rhenium decreased carbon deposition even further, and the life of a reforming catalyst is now measured in years, whereas the life of a cracking catalyst was measured in hours. The patent literature now abounds with the names of other elements whose compounds are claimed as promoters, or retarders of carbon formation: these include most frequently lead, tin, and germanium.

Dual-function catalysts have some role to play in the cracking of high molecular weight hydrocarbons, but this is not their primary function. Although it is economically desirable to perform as many operations as possible in one reactor, considerations of catalyst design nevertheless make it preferable to have different catalysts for different processes, and indeed for different types of feedstock. Thus for example the operating conditions for naphtha reforming (750 to 820 K, 20 to 40 atm—2 to 4 MPa) are more stringent than those required to isomerize linear C_4 to C_6 hydrocarbons, and the catalysts used in the two processes differ considerably in composition.

Let us return now to consider a little further the reaction pathways which molecules follow during catalytic reforming. Figure 7.1 illustrates the transformations possible with n-hexane, reactions on the metal (involving addition

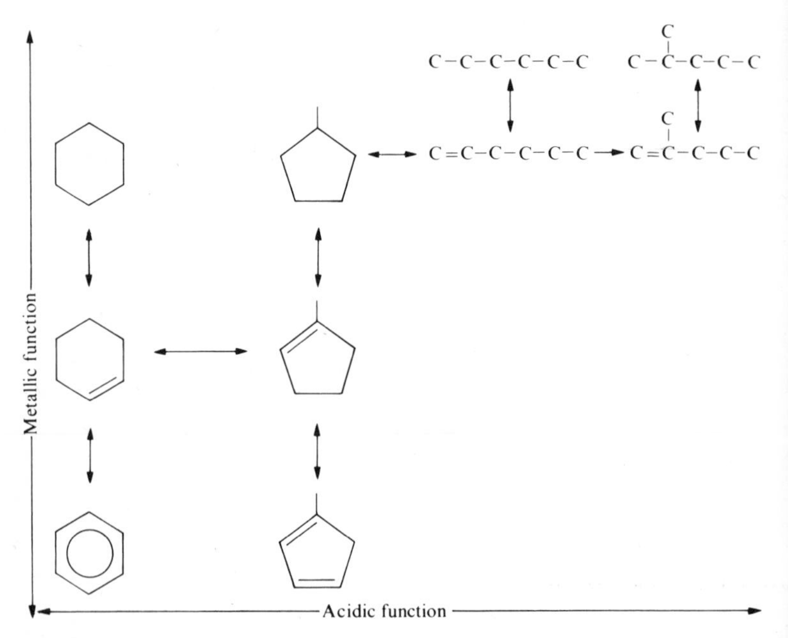

FIG. 7.1. Pathways for reaction of n-hexane on a dual-function reforming catalyst.

or abstraction of hydrogen) occurring vertically, and those on the acidic support (involving isomerization or ring closure) being shown horizontally. According to this scheme, C_6 rings are not formed directly from a linear C_6 olefin, but necessitate passing through methylcyclopentane. While there is much evidence for this, there is also evidence for direct C_6 ring formation, but this appears to be able to proceed on metallic sites by a non-ionic mechanism.

It has been possible to offer a convincing demonstration that olefinic intermediates migrate through the gas phase from one type of site to another. By mixing finely divided platinum/silica (which has no acidic attributes) with acidic alumina (which of course has no metallic properties), conversions have been observed which approximate to those found when platinum is supported directly on acidic alumina. Close proximity of the two kinds of site is therefore not essential.

One final point for clarification concerns reactions on the metallic function. It has been stated above that aromatics are desirable products for their high octane ratings, and that dehydrogenation of aliphatic hydrocarbons occurs before carbonium ions can be formed: and yet these reactions are conducted in the presence of hydrogen, which seems entirely the wrong thing to do. The answer to this dilemma is provided by thermodynamics. As soon as cyclohexane is formed from methylcyclopentene (Fig. 7.1), it will at temperatures above about 570 K be substantially converted to benzene, even when excess hydrogen is present. Cyclohexane is not a principal product of catalytic reforming, since ΔG for its dehydrogenation to benzene is zero at about 470 K. The free-energy change for dehydrogenation of aliphatic hydrocarbons is much less favourable, but calculation shows that reforming reactions can be sustained even if the concentration of olefin in the gas phase is as low as 10^{-7} atm (10^{-2} Pa).

8. Catalysis in the petrochemical industry

8.1. Objectives of the industry

WITHIN the lifetime of readers of this book, the heavy organic chemicals industry has changed from being based essentially on coal and fermentation processes to being almost completely oil-based. In 1949, oil was the raw material for less than 10 per cent of the organic chemicals made in this country; by 1963 it had risen to over 60 per cent, and the figure now must be even higher. The reasons for this change are complex, but are all basically economic: in comparison with coal, oil is more easily and cheaply transported (e.g. by pumping through pipelines), its extraction from the ground is less labour-intensive and therefore also cheaper, and because of these factors it suffers less from short-term fluctuations.

In the modern petrochemical industry, products derived from crude oil are converted by the skill of chemists into a multiplicity of useful products, chief amongst which are plastics, elastomers, and synthetic fibres. It is easy to forget that before the Second World War the only synthetic fibre available was Rayon, and the only plastics widely used were condensation polymers of the phenol-formaldehyde type: there were no detergents as we know them, no Nylon and no polythene. Those who would criticize the chemical industry because of the pollution it creates would do well to ask themselves whether they would really like to be without its products.

The vast majority of synthetic products that make life easier and more enjoyable have thus been developed within the last two or three decades. Their availability at a price which ensures their being widely used is a direct consequence of the change from carbohydrates and coal to oil as the basic raw material. Almost all of the organic products of the petrochemical industry are made from olefins, especially ethylene, propylene, and the butenes: the richness and diversity of their chemistry enables them to be polymerized, co-polymerized, and selectively oxidized to a myriad of useful products. It is because they can be made on a large scale, and therefore cheaply, that the petrochemical industry has grown as it has done.

We will now examine very briefly how olefins are made, and then review how catalytic processes contribute to the operation of the petrochemical industry.

8.2. The manufacture of olefins

This subject will only be given the most superficial of treatments because the principal processes used are non-catalytic; some mention of it is however necessary for completeness. In the United Kingdom and Western Europe, some 80 per cent of all the ethylene manufactured comes from the thermal cracking of the fraction of crude oil described as 'naphtha' (see Chapter 7). The greater the severity of the reaction conditions, i.e. the higher the temperature and the longer the contact time, the lower is the average number of carbon atoms in the product. In the region of 1100 K, there is about twice as much ethylene as propylene; the use of higher temperatures yields more ethylene and less propylene and butenes, and vice versa. The state of hydrogenation of the products is also temperature-dependent because of differences in thermodynamic stability. The free energies of formation of ethane, ethylene, and acetylene are such that below 1125 K ethane is the most stable, while above 1400 K acetylene is the stablest. Methane and hydrogen are also substantial products of thermal cracking, and traces of higher hydrocarbons are always present. Steam is added to the naphtha vapour to decrease its partial pressure and thus to encourage dehydrogenation; it also removes carbon deposits. Contact times are of the order of one second, and the products are either quenched to preserve the higher temperature concentrations or cooled slowly to give the products which are stable at lower temperatures. Because of economy of scale, plants are now built to handle 1500 tons per day (or $0{\cdot}5 \times 10^6$ tons per year) of naphtha.

In the United States, olefins are predominantly obtained by dehydrogenation of the saturated hydrocarbons higher than methane present in 'natural gas'.

8.3. Reactions of ethylene

The principal reactions of ethylene employed in the petrochemical industry are displayed in Fig. 8.1; not all useful reactions are shown, since some of these are non-catalysed. They fall into four categories.

(1) Reactions involving addition across the double bond (e.g. hydration of ethylene to ethanol).
(2) Reactions involving substitution of a hydrogen atom, sometimes followed by rearrangement (e.g. ethylene to vinyl chloride).
(3) Insertion reactions in which the double-bond character is lost (e.g. oxidation of ethylene to ethylene oxide).
(4) Polymerization reactions.

It is a matter of nice definition whether the latter should be considered as catalysed processes, since frequently the 'catalyst' acts only as an initiator of a chain reaction; however in the case of stereospecific polymerizations occurring in a heterophase system, the catalyst plays a vital role in determining the nature of the product. We cannot however give an adequate treatment of this fascinating branch of chemistry here, and the matter must be left at this point.

The most important way in which catalysis enters into the useful reactions of ethylene is in its partial oxidation. The cost of converting ethylene to an oxidized product is greatly dependent on the number of steps in the process,

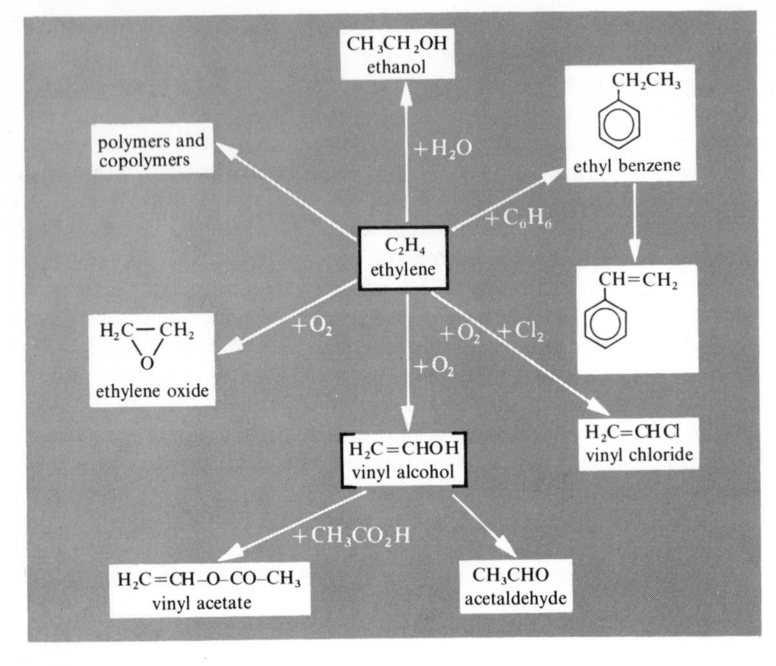

FIG. 8.1. Some industrially important reactions of ethylene.

and the science of catalysis has been fruitfully applied, especially in the past decade, to achieving simple one-step conversions which have rapidly replaced more lengthy and therefore more expensive routes. A good example of this is the manufacture of ethylene oxide. This was formerly done by a two-stage 'chlorohydrin' process involving first the addition of hypochlorous acid (chloric (I) acid), followed by elimination of hydrogen chloride:

$$CH_2{=}CH_2 + HOCl \rightarrow CH_2Cl{-}CH_2OH \rightarrow \underset{\diagdown O \diagup}{CH_2{-}CH_2} + HCl.$$

This route has now been almost entirely replaced by the one-step direct oxidation using a silver catalyst at about 520 K:

$$CH_2{=}CH_2 + \tfrac{1}{2}O_2 \rightarrow \underset{\diagdown O \diagup}{CH_2{-}CH_2}.$$

Ethylene oxide can be obtained in about 70 per cent yield, the remainder of the ethylene appearing as carbon dioxide. The silver catalyst is promoted to achieve this selectivity, one way of doing this being to add continuously a small amount of a chlorinated hydrocarbon to the reactants. This gives a partial coverage of the surface by chloride ions; this in turn inhibits the dissociative adsorption of oxygen, which is beneficial because the selective oxidation is known to require adsorbed oxygen *molecules.*

Two other examples will suffice to show the power of selective oxidation catalysis. Acetaldehyde was formerly made by dehydrogenation of ethanol. During the late 1950s a homogeneous catalytic route known as the Wacker process was developed, and has since been widely employed. The mechanism need not concern us, but it led to the discovery that the reaction could also be performed heterogeneously by passing ethylene and oxygen over a supported palladium catalyst at about 420 K:

$$CH_2{=}CH_2 + \tfrac{1}{2}O_2 \rightarrow CH_3CHO.$$

The unstable vinyl alcohol is probably formed first, and for this reason the reaction falls into the second of the categories listed above. This has in turn led to a much simplified route to vinyl acetate, a valuable monomer which can be converted to polyvinyl acetate and polyvinyl alcohol. Previously made by addition of acetic acid to acetylene, viz.

$$CH{\equiv}CH + CH_3CO_2H \rightarrow CH_2{=}CH{\cdot}O{\cdot}COCH_3$$

it is now made in high yield simply by passing ethylene, oxygen, and acetic acid over a palladium catalyst:

$$CH_2{=}CH_2 + CH_3CO_2H + \tfrac{1}{2}O_2 \rightarrow CH_2{=}CH{\cdot}O{\cdot}COCH_3 + H_2O.$$

The last example is the synthesis of vinyl chloride. This used to be made either by adding hydrogen chloride to acetylene, or by chlorinating ethylene and then dehydrochlorinating the product, viz.

$$CH_2{=}CH_2 + Cl_2 \rightarrow CH_2Cl{-}CH_2Cl \rightarrow CH_2{=}CHCl + HCl.$$

For the process to be economic, the hydrogen chloride had then to be oxidized back to chlorine by the old Deacon process using a copper catalyst. It has recently been found that a one-step *oxychlorination* process can be effected over copper(II) chloride at 420 K:

$$2CH_2{=}CH_2 + Cl_2 + \tfrac{1}{2}O_2 \rightarrow 2CH_2{=}CHCl + H_2O.$$

A point that emerges repeatedly from the foregoing examples is that processes based on ethylene have tended to replace those starting with acetylene. There are two quite independent reasons for this. The first is simply that ethylene is cheaper than acetylene, now that the former is made by cracking of naphtha. The second reason is a technical one: addition reactions of acetylene are all quite exothermic and therefore not easy to control, whereas reactions of ethylene are much better tempered in this respect.

8.4. Reactions of propylene

The chemistry of propylene is more extensive than that of ethylene, for, in addition to the classes of reaction listed for the latter at the beginning of the last section, the propylene molecule can also be oxidized at the methyl group. Some discussion of the selective oxidation of propylene has already been presented (section 2.8). There it was noted that selective oxidation catalysts of the 'bismuth molybdate' type attack the molecule at the methyl group, since the first stage of its activation is through loss of a hydrogen atom to give an allyl radical. It is interesting to note that, just as palladium catalyses the oxidation of ethylene to acetaldehyde, so it can also oxidize propylene to acetone. Clearly the metal-catalysed and oxide-catalysed oxidations proceed by quite different mechanisms.

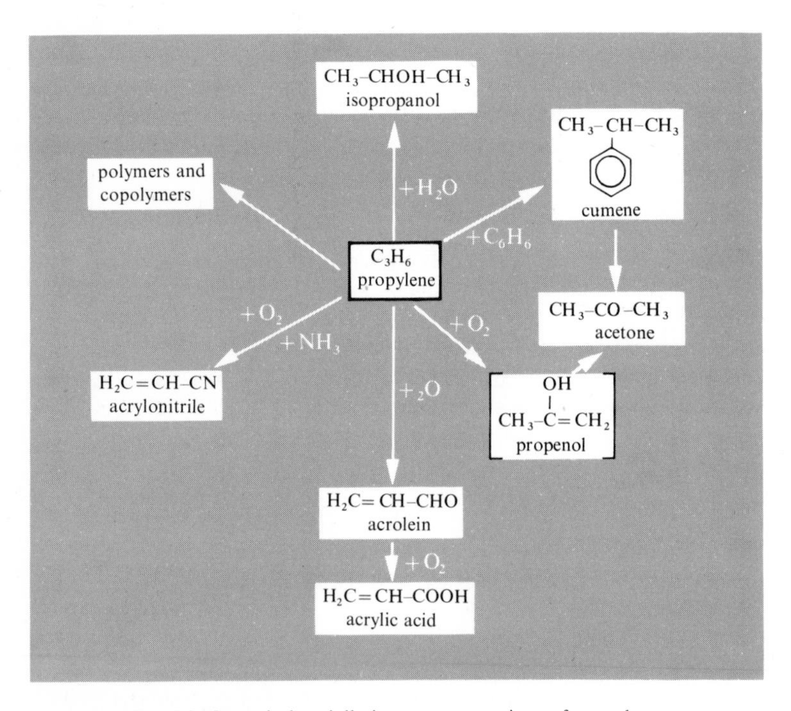

FIG. 8.2. Some industrially important reactions of propylene.

Some of the more industrially important reactions of propylene are presented in Fig. 8.2. There are of course some similarities to ethylene, for example, in polymerization. The hydroformylation of olefins has not been mentioned previously, and a word is in order now. An olefin can react with hydrogen and carbon monoxide in the presence of a suitable catalyst to give an aldehyde or an alcohol, e.g.

$$CH_3{-}CH{=}CH_2 + H_2 + CO \rightarrow CH_3{-}CH_2{-}CH_2{-}CHO$$

or

$$CH_3{-}CH{=}CH_2 + 2H_2 + CO \rightarrow CH_3{-}CH_2{-}CH_2{-}CH_2OH.$$

The usual catalyst is dicobalt octacarbonyl, $Co_2(CO)_8$, or the derived hydride, $HCo(CO)_4$: this is used in solution, so the reaction is really homogeneously catalysed. Secondary aldehydes and alcohols are also produced, but these are less useful.

One important point of difference between the two olefins is that propylene oxide cannot be made by the silver-catalysed oxidation route: it still has to be made by the 'chlorohydrin' process which is now outmoded for ethylene.

We should perhaps say a little more about the important oxidation of propylene to acrolein and its even more important ammoxidation to acrylonitrile (see p. 34). The latter reaction almost certainly involves the formation of acrolein as an intermediate, this point being the basis of an epic patent law suit. Acrolein is a most toxic substance and is difficult to handle; it is employed in the synthesis of glycerol, but otherwise is not yet widely used. Acrylonitrile on the other hand when copolymerized with a basic monomer such as 2-vinylpyridine forms a valuable acrylic fibre (e.g. Acrilan), and is a component together with butadiene (section 8.5) and styrene of the polymeric matrix employed as synthetic rubber. A further important use is mentioned later on (section 8.6). The ammoxidation is performed at 670 to 770 K, and the selectivity to acrylonitrile exceeds 75 per cent: other products are hydrogen cyanide, now produced in sufficient amount to render its synthesis by other routes unnecessary, and acetonitrile, for which there is an inadequate use.

8.5. Reactions of butenes

Many of the reactions of the butenes are analogous to those of ethylene and propylene, and so need not detain us. There are however two important reactions specific to the n-butenes; these are (i) their dehydrogenation to 1,4-butadiene and (ii) their oxidation to maleic anhydride.

Butadiene is manufactured on a very large scale, and is principally employed to create elastomers by polymerization or copolymerization with other monomers. In Europe enough can be obtained as a by-product of naphtha cracking under the appropriate conditions, but especially in the United States it is made by dehydrogenation of n-butenes or butane. The reaction, which is endothermic, is conducted on a chromia–alumina catalyst rather than on a metal catalyst, for although coke formation occurs on both the former is regenerable with less damage to itself. For endothermic reactions, the temperature falls on passing along the catalyst bed which is therefore usually kept broad and shallow to minimize this effect. The temperature used for this dehydrogenation is about 870 K and the pressure approximately 3 atm (300 kPa). The reactant is diluted with a large excess of steam which keeps the partial pressure of reactant low, and encourages the reaction to proceed in the forward direction.

It is also possible to perform an oxidative dehydrogenation, i.e.

$$C_4H_8 + \tfrac{1}{2}O_2 \rightarrow C_4H_6 + H_2O$$

using catalysts of the 'bismuth molybdate' type (see Table 2.5). The position of equilibrium here is much more favourable to obtaining high yields of butadiene than in the simple dehydrogenation system.

The n-butenes can be selectively oxidized to maleic anhydride in the presence of divanadium pentoxide (see Table 2.5), although the process is not operated commercially in this country.

8.6. Reactions of aromatic hydrocarbons

The catalytic chemistry of aromatic hydrocarbons as practised on a large scale is largely associated with the production of synthetic fibres, especially Nylon-6 (made by high temperature polymerization of caprolactam) and Nylon-6,6 (made by condensation polymerization of adipic acid and hexamethylene diamine). These reactions are shown in Fig. 8.3.

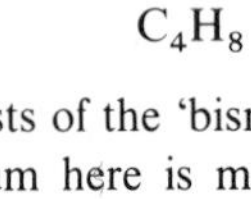

$$\text{caprolactam} \longrightarrow \left[NH{-}(CH_2)_5{-}CO{-}NH{-}(CH_2)_5{-}CO\right]_n$$

caprolactam — Nylon–6 polymer

$$HOOC{-}(CH_2)_4{-}COOH + H_2N{-}(CH_2)_6{-}NH_2 \longrightarrow \left[CO{-}(CH_2)_4{-}CO{-}NH{-}(CH_2)_6{-}NH\right]_n$$

adipic acid — hexamethylene diamine — Nylon–6,6 polymer

FIG. 8.3. Synthesis of Nylon-6 and Nylon-6,6.

Three distinct routes to caprolactam have been used (see Fig. 8.4), all starting from benzene and all leading to cyclohexanone oxime, which can undergo an acid-catalysed Beckmann rearrangement to form caprolactam. In the route developed first, benzene was alkylated with propylene to give isopropylbenzene (cumene), which was then oxidized to cumene hydroperoxide which readily undergoes an acid-catalysed reaction to phenol and acetone. This reaction is still of course used to make these products, which

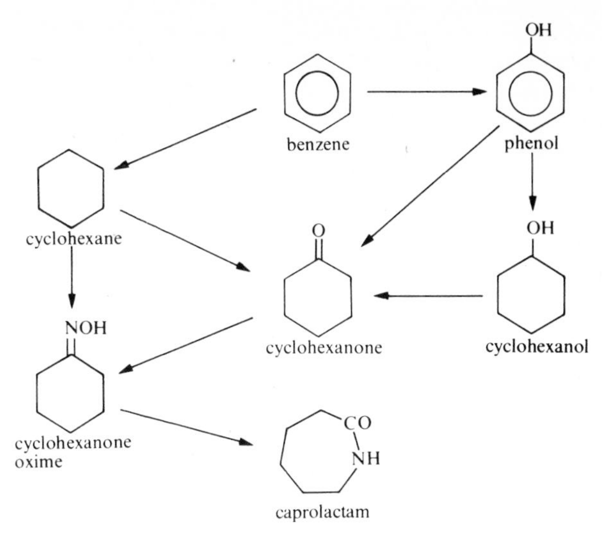

FIG. 8.4. Routes to caprolactam.

have many other applications. It is interesting to note that despite the skill and effort of catalytic chemists, there is still no process for the one-stage oxidation of benzene to phenol. Phenol was then hydrogenated to cyclohexanol and dehydrogenated to cyclohexanone, from which the oxime is easily made. In a new variant on this route, phenol is hydrogenated directly to cyclohexanone using a dual-function palladium/alumina catalyst. The first product is the unstable cyclohexenol, which then rearranges in an acid-catalysed reaction on the alumina to give cyclohexanone:

$$\text{phenol} \xrightarrow{+2H_2} \text{cyclohexenol} \rightarrow \text{cyclohexanone}$$

The two routes currently preferred for making cyclohexanone oxime both start by hydrogenating benzene to cyclohexane, using a nickel catalyst at a moderate temperature. In one route, this then undergoes 'photonitrosation' to give the oxime directly, or in the other route it is oxidized by air in the presence of a soluble cobalt catalyst to cyclohexanone.

Cyclohexane is also used to make adipic acid for Nylon-6,6, but the processes involved are non-catalytic. Adipic acid is also an intermediate in one route to hexamethylene diamine; it is converted to the amide with ammonia, dehydrated to adiponitrile, and then hydrogenated. An interesting and much shorter route is based on acrylonitrile made by ammoxidation of propylene (see pp. 34 and 93) using 'bismuth molybdate' or a similar catalyst; acrylonitrile is then dimerized in an electrochemical hydrogenation to give hexamethylene diamine in one step.

The direct catalytic oxidation of aromatic hydrocarbons to phthalic anhydride is also practised on a large scale. Either *o*-xylene or naphthalene may be used, and the catalyst is divanadium pentoxide (see Table 2.5): short contact times at about 670 K are used, and efficient removal of heat of reaction is extremely important.

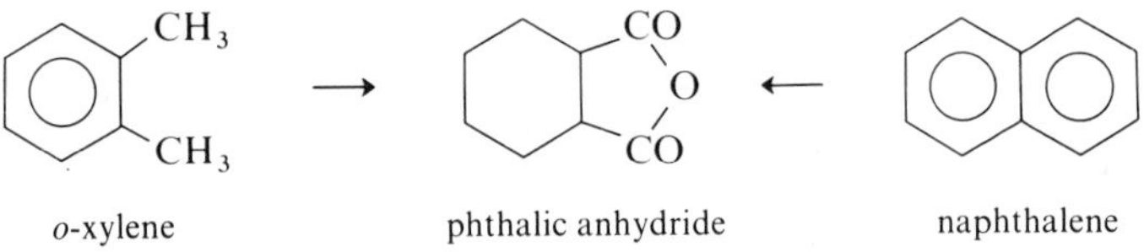

CH_3 CH_3 → CO O CO ←

o-xylene phthalic anhydride naphthalene

The principal use of phthalic anhydride is to make long-chain esters (e.g. dinonyl phthalate) used as plasticizers for poly(vinyl chloride).

Styrene is manufactured on a large scale by alkylation of benzene with ethylene to give ethylbenzene, followed by dehydrogenation under conditions similar to those used for converting butenes to butadiene (see section 8.5).

Numerous other catalytic operations are conducted on products derived from aromatic hydrocarbons. One example of this is palladium- or platinum-catalysed hydrogenation of dinitrotoluenes to tolylene diamines used as intermediates in making tolylene di-isocyanates: these form the basis of polyurethane foams.

QUESTIONS

8.1. Why do you think silver is much less willing to catalyse the oxidation of propylene to propylene oxide than that of ethylene to ethylene oxide?

8.2. How may acrolein be converted to glycerol?

8.3. Explain why you think higher equilibrium yields of butadiene can be obtained by oxidative dehydrogenation of butenes than by simple dehydrogenation.

8.4. Why is it so difficult to oxidize benzene to phenol?

8.5. Try to find out more about the electrochemical hydrogenation and dimerization of acrylonitrile to hexamethylene diamine.

9. Steam-reforming of hydrocarbons

Production and uses of 'synthesis gas'

9.1. Introduction

MANY of the important industrial processes which have already been mentioned require hydrogen as one of the reactants: fat hardening and petroleum reforming are two examples. A further large scale process utilizing hydrogen is ammonia synthesis, and this will be treated in the next chapter. For many decades industrial scientists have therefore sought means for the cheap and efficient manufacture of hydrogen. We now consider briefly how this used to be done.

The manufacture of pure hydrogen on a large scale cannot be performed in one step, except by electrolysis of water, which is economic only where cheap electricity (e.g. hydro-electric generation) is available. Instead it is usual to make '*synthesis gas*', which is a generic name for mixtures of hydrogen, carbon monoxide, and carbon dioxide in various proportions: this may then be treated to give pure hydrogen, or alternatively used directly in the applications described below. The old method of making synthesis gas involved the *water gas reaction*. Here steam was passed over red-hot coke, when the following endothermic reactions occurred:

$$C + H_2O \rightleftharpoons H_2 + CO$$

$$C + 2H_2O \rightleftharpoons 2H_2 + CO_2$$

When the coke had cooled to a point where the reaction was too slow to be useful, air was introduced to re-heat the bed of coke by the following exothermic reactions:

$$C + \tfrac{1}{2}O_2 \rightleftharpoons CO$$

$$C + O_2 \rightleftharpoons CO_2$$

Steam was then re-introduced, and the reaction continued in this cyclical manner.

Of the many unattractive features of this process, the one needing emphasis is that the hydrogen produced came solely from the water. It has long been recognized that hydrocarbon feedstocks are quite rich in hydrogen, the hydrogen-to-carbon ratio varying from four in the case of methane to about two for long-chain hydrocarbons. Thus for example using methane in place of coke, two additional molecules of hydrogen are formed in the reaction with steam:

$$CH_4 + H_2O \rightarrow 3H_2 + CO$$

As methane and other hydrocarbon sources have become cheaper and more readily available, the production of hydrogen and of synthesis gas by the *steam-reforming* of hydrocarbons has supplanted the water gas reaction, and now some 90 per cent of the world's hydrogen production is performed by this newer route.

9.2. Steam-reforming of hydrocarbons

The type of hydrocarbon used in the steam-reforming process depends upon local availability. Thus in Italy and the United Kingdom, 'natural gas' (predominantly methane) is used, whereas, in other countries not so adequately endowed with it, fuel oil or naphtha are employed instead. The primary product of the reaction of methane with steam is a mixture containing hydrogen, carbon monoxide and dioxide, and unchanged reactants, but the *water gas shift equilibrium*

$$CO + H_2O \rightleftharpoons CO_2 + H_2$$

is established thereafter. The methane–steam reaction is endothermic ($\Delta H = +206\ \text{kJ mol}^{-1}$), and is conducted in a multi-tubular reactor (section 4.3) at 1020 to 1120 K, depending on the required product composition.

Metallic nickel is the active catalytic species in all commercial steam-reforming catalysts. The noble Group VIII metals and cobalt are also effective, but are not used because of their higher cost. The operating conditions are severe, and loss of metal area by sintering is a real danger: much attention has therefore been given to the selection of suitable supports that are stable for prolonged periods, and of other additives designed to inhibit sintering and to ensure mechanical strength and stability. Suitable supports are α-alumina and magnesia, or mixtures thereof. Physical strength is obtained by incorporating an hydraulic cement among the ingredients. A calcium aluminate type is preferred to a siliceous Portland cement, for under operating conditions silica is volatile as 'orthosilicic acid', $Si(OH)_4$, and this leads to progressive structural weakness. ICI manufactures three different catalysts for steam-reforming; catalyst 57-1 is recommended for primary reforming of natural gas and methane at 30 atm (3 MPa) and 1020 K, while 46-1 is suitable for the reforming of higher hydrocarbons under these conditions. For the preparation of a nitrogen–hydrogen mixture for ammonia synthesis, air is added to the product of the primary reforming: the temperature is thereby raised, and the gases pass through a secondary catalytic reformer at about 1570 K, giving a product which has only a low methane content. The ICI catalyst 54-2 is successful under these very demanding conditions. The compositions of these catalysts are listed in Table 9.1.

In common with other metallic catalysts, nickel steam-reforming catalysts are readily poisoned by the sulphur compounds present in natural gas and naphtha. It is therefore necessary to pretreat the feedstock to eliminate them,

TABLE 9.1

Composition of ICI steam-reforming catalysts

	Composition, %		
	57-1	46-1	54-2
NiO	32	21	18
CaO	14	11	15
SiO_2	0·1	16	0·1
Al_2O_3	54	32	67
MgO	—	—	—
K_2O	—	7	—

and this is done in two ways. Zinc oxide is used to remove reactive sulphur compounds (e.g. mercaptans and thiols) in a non-catalytic manner, the oxide being converted to sulphide in the process. Non-reactive aromatic sulphur compounds (e.g. thiophene) are removed by hydrodesulphurization using 'cobalt molybdate' as catalyst: hydrogen has to be introduced for this purpose, and the products from thiophene are hydrogen sulphide (removed by further treatment with zinc oxide) and a C_4 hydrocarbon.

Carbon deposition is a further serious problem encountered in steam-reforming. It occurs through decomposition of carbon monoxide, or its reduction by hydrogen, and by dehydrogenation of the hydrocarbon feedstock. Aromatic hydrocarbons present in naphtha are particularly vicious depositors of carbon. It can be removed by reaction with steam or carbon dioxide, but treatment with hydrogen is comparatively ineffective.

Mention has been made above of the water gas shift equilibrium. When hydrogen is the desired product of steam-reforming, as for example when it is required for ammonia synthesis or a catalytic hydrogenation, carbon monoxide may be removed, or at least reduced in concentration, by reaction with water. The process is operated in a 'high temperature' (about 770 K) or a 'low temperature' (about 500 K) regime. The former is generally used when hydrogen is needed for ammonia synthesis, but an absorption process to remove residual carbon monoxide has to follow. The 'high temperature shift' catalyst contains magnetite (Fe_3O_4) and chromia (Cr_2O_3). ICI have recently developed a 'low temperature shift' catalyst 52-1 containing copper(II) oxide (30%), zinc oxide (45%), and alumina (13%): the copper oxide is reduced to metallic copper *in situ*. The advantage of working at lower temperature with a more active catalyst is that the equilibrium yield of carbon monoxide is lower, and whatever is left can be removed by methanation (reaction with hydrogen over a nickel catalyst to give methane), thus avoiding the need for costly absorption equipment.

9.3. Uses of synthesis gas

In Britain one of the principal uses of synthesis gas made by the steam-reforming of hydrocarbons is as a fuel (town gas) for domestic and industrial heating. In the two decades between 1950 and 1970, the manufacture of town gas was switched almost entirely from coal distillation to steam-reforming: subsequently of course it has been rapidly replaced by North Sea gas. One of the consequences of the earlier change was a severe shortage of smokeless fuels, and of some organic chemicals, previously products of coal distillation. The process continues to be applied in areas where natural gas is not abundantly available.

A second important application of synthesis gas is for the manufacture of methanol according to the equation

$$CO + 2H_2 \rightarrow CH_3OH$$

Carbon dioxide may also be used, but it is first reduced by hydrogen to carbon monoxide which then reacts as above. This reaction is exothermic ($\Delta H = -91\ \text{kJ mol}^{-1}$) and the equilibrium methanol yield therefore decreases with increasing temperature. The use of an active catalyst is essential to obtaining high yields per pass (see section 1.1) and of course high pressure is also helpful. For many years the catalyst was based on zinc oxide with chromia: it operated at 570 to 670 K and 300 atm (30 MPa) pressure, and conversions of 12 to 15 per cent per pass were obtained. In the mid-1960s however an improved catalyst using metallic copper was introduced; this works under much milder conditions (550 K, less than 50 atm—5 MPa), and so plants using this new catalyst are much cheaper to build.

Methanol is the starting material for a number of useful organic products. Much is converted by catalytic processes to formaldehyde, either by oxidative dehydrogenation

$$CH_3OH + \tfrac{1}{2}O_2 \rightarrow HCHO + H_2O$$

or by direct dehydrogenation

$$CH_3OH \rightarrow HCHO + H_2.$$

The former reaction is catalysed with good selectivity by the oxides of vanadium and molybdenum, but it is quite exothermic ($\Delta H = -159\ \text{kJ mol}^{-1}$) and heat removal is therefore a problem. Silver catalyses both reactions, and since the latter is endothermic ($\Delta H = +84\ \text{kJ mol}^{-1}$) temperature control becomes easier. Silver is used either in the form of a bed of gauzes or as granular particles of carefully graded size. The principal uses of formaldehyde are to make cross-linking agents for thermosetting resins (e.g. pentaerythritol and hexamethylene tetramine), and in phenol–formaldehyde and urea–formaldehyde resins.

One further application of synthesis gas which must be mentioned is the Fischer–Tropsch synthesis. In this reaction hydrogen and carbon monoxide react catalytically to give higher-molecular-weight products, which may be hydrocarbons when cobalt is the catalyst and various straight-chain acids, aldehydes, and alcohols when iron is used. The reaction over ruthenium gives solid polymers of very high molecular weight. Although much research has been done on these reactions, the hydrocarbon products which might be used as fuels can be made economically only when other sources are absent, and indeed the synthesis gas would in this case be derived from coal. The only country in which the application of the Fischer–Tropsch synthesis appears to have been attempted is South Africa.

Synthesis gas is also used for the hydroformylation of olefins (see section 8.1). Acetic acid is now made on a large scale by the carbonylation of methanol according to the equation

$$CH_3OH + CO \rightarrow CH_3CO_2H$$

by a homogeneously catalysed reaction in which rhodium tri-iodide dissolved in methanol acts as the catalyst.

Several other useful products arise as by-products of synthesis gas manufacture; these include argon and carbon dioxide.

QUESTIONS

9.1. On the basis of the ideas presented in sections 2.5 and 2.6, defend the selection of copper as a catalyst for the water gas shift and for the synthesis of methanol.

9.2. Outline possible synthetic routes to vinyl acetate, starting from air, synthesis gas, and ethylene. What catalysts might be used?

10. Catalysis in the heavy inorganic chemicals industry

10.1. Introduction

THERE are several large-scale industrial processes which are conveniently treated together, not so much because of their similarity but because the application of their products is interlinked, and also because they are all inorganic reactions. The three reactions to be considered in this chapter are (i) the synthesis of ammonia, (ii) its oxidation to oxides of nitrogen from which nitric acid is made, and (iii) the oxidation of sulphur dioxide to the trioxide for the manufacture of sulphuric acid.

Ammonia is manufactured from a mixture of nitrogen and hydrogen (prepared as described in the last chapter) on a very large scale indeed; in 1970 the U.S.A. produced about 2×10^7 tons and Western Europe about $1{\cdot}2 \times 10^7$ tons. Approximately one-quarter of this is used to make nitric acid, and about 70 per cent of it is used in the making of fertilisers. The principal inorganic fertilisers are ammonium nitrate and ammonium sulphate, and so the close links between the three reactions we are now to consider becomes evident. The use of urea ($H_2N{\cdot}CO{\cdot}NH_2$) as a fertilizer is growing rapidly: it is made from ammonia and carbon dioxide, but the reaction does not require a catalyst.

10.2. The synthesis of ammonia

The history of the development of this classic catalysed reaction is replete with cautionary tales which demonstrate the happy blend of science and good luck necessary to evolve a major industrial process. The very early attempts to effect the combination of nitrogen with hydrogen by non-catalytic means, made towards the end of the last century, were foredoomed to fail because the essential thermodynamics of the system were not known. The work of Haber and his associates in the first decade of this century, since extended and refined by many others, showed that the reaction

$$N_2 + 3H_2 \rightarrow 2NH_3$$

was exothermic ($\Delta H = -46\,\mathrm{kJ\,(mol\,NH_3)^{-1}}$), so that highest equilibrium yields of product will be obtained at low temperatures and high pressures. Here is another example of a situation where kinetic and thermodynamic considerations are in conflict, and there is a clear need for an active and stable catalyst to bring the system to equilibrium at the lowest possible temperature.

The search for such a catalyst was conducted industriously and patiently by Haber, Bosch, and Mittasch in the laboratories of BASF (Badische Anilin und Soda Fabrik) between about 1905 and 1910: it is a story of frustration and ultimate triumph. It was soon established that a number of metals, including tungsten, uranium, iron, ruthenium, and osmium, had some activity, but unfortunately it was short-lived. Part of the programme consisted of testing iron ores from various sources; these were then reduced to give metallic iron catalysts. Success came when a catalyst made from an ore obtained from Galliväre in Sweden was tested, and found to show satisfactory stability: analysis showed it to contain some alumina and potassium oxide, and the way was immediately open for the development of the promoted iron catalyst which is still used with only minor modifications today. Large-scale production of synthetic ammonia using such a catalyst started in Germany in 1914.

The catalyst is now made by fusing magnetite with the necessary amounts of promoters, and, after solidification of the melt, the material is crushed and screened to get the desired particle size. The reason for the continued use of magnetite is interesting. It has a spinel structure consisting of a cubic packing of oxide ions in whose interstices ferrous and ferric ions are alternately located (see Fyfe: *Geochemistry* (OCS 16)). On reduction, which may be performed either in the plant or externally, no relaxation of the lattice occurs, and so removal of the oxide ions leaves a porous iron matrix having a surface area of some 20 m^2g^{-1}.

The promoters play a vital but complex part in ensuring long life for the catalyst. ICI catalyst 35-4 contains 0·8% potassium oxide, 2·0% calcium oxide, 2·5% alumina, 0·4% silica, 0·3% magnesia, and traces of the oxides of titanium, zirconium, and vanadium. The essential role of the alumina is to stabilize the iron crystallites against sintering. It covers much of the iron surface and acts as a barrier to the fusion of the small iron particles: it is termed a *structural promoter*. Potassium oxide plays a different part: it increases the activity of the iron surface, and is referred to as an *electronic promoter*.

The process is usually operated at 150 to 350 atm (15 to 35 MPa), the equilibrium conversion to ammonia being about 50 per cent at 670 K. Much higher pressures (about 1000 atm) are sometimes used, but the advantage of higher conversions scarcely offsets the greatly increased capital investment in plant.

The kinetics of the reaction have been investigated by numerous workers, and the mechanism of the reaction is by no means unequivocally established. It was believed for a long time that the adsorption of nitrogen was the rate-limiting step, but more recently the possibility that the reaction between an adsorbed nitrogen molecule and adsorbed hydrogen is the slow step has been considered. The kinetic equations derived to describe the rate are complex due to the necessity of making allowance for the reverse reaction.

The iron catalyst, like most metallic catalysts, is extremely sensitive to poisons which include oxygen-containing compounds (e.g. water, carbon monoxide and dioxide) which cause partial oxidation of the surface, and sulphur compounds which cause sulphiding: 0·01% of sulphur in the feed gas has an observable effect on activity. The effect of oxygen compounds is reversible provided that they have not been present for more than a few days.

10.3. The oxidation of ammonia

The selective oxidation of ammonia to nitric oxide is accomplished by quickly passing a mixture of the reactants through a metallic catalyst at temperatures above about 1020 K: nitric oxide is subsequently oxidized to nitrogen dioxide which is then absorbed in water. The reactions involved are

$$4NH_3 + 5O_2 \rightarrow 4NO + 6H_2O$$

$$2NO + O_2 \rightarrow 2NO_2$$

$$3NO_2 + H_2O \rightarrow 2HNO_3 + NO$$

Nitric oxide formed in the last reaction is further oxidized and is in turn converted to nitric acid.

The reaction has a long history. The first patent was obtained by Kuhlmann in 1838; the catalyst employed by him was platinum sponge. Industrial application of the process had to await the availability of high-purity ammonia, since that made from gas-works liquors contained enough sulphur and arsenic compounds to give rapid poisoning of the catalyst. The reaction occurs rapidly and contact times of 10^{-3} to 10^{-4} second are enough to give high conversions. To facilitate rapid passage of the gases, the catalyst form which has long been in use is a bed of finely woven gauzes of a platinum alloy. It is remarkable that the gauze dimensions have altered scarcely at all since this form of catalyst was first introduced by Kaiser in 1909: the wire diameter is 0·06 mm, and there are 1050 apertures per cm^2. The oxidation of ammonia to nitric oxide is extremely exothermic ($\Delta H = -900$ kJ mol^{-1}) and provided the incoming gas is heated to about 550 K the heat of reaction is large enough to maintain the catalyst bed at the desired temperature.

The process is operated under a range of pressure conditions from atmospheric to about 8 atm (800 kPa). Low-pressure reactors are large, with gauze diameters up to 4 m; the high pressure reactors are smaller, with gauze diameters of about 1 m, and operating temperatures are high (1170 to 1220 K). Gauzes rest on ceramic supports, and the gas stream flows downwards to prevent the gauze flapping. The selectivity to nitric oxide is typically 95 to 97 per cent, the remainder of the ammonia being converted to nitrogen, probably by decomposition on the hot walls of the reactor. The reactant stream contains 10 to 12 per cent ammonia, and secondary air is added to effect the oxidation to nitrogen dioxide, absorption of which in water yields nitric acid of about 60 per cent strength.

A peculiar feature of the reaction is the slow physical changes which gauzes undergo while on stream. Initially starting as a surface roughening, they progress to form excrescences, some of which are blown off by the rapid flow of gas, and extensive crystal growth occurs. As a result of these processes, the gauzes gradually weaken and ultimately break and have to be replaced. The cause of the effects is uncertain, but it is possible that the surface temperature is much higher than the average temperature, and that the surface atoms are highly mobile. The deterioration is less rapid with platinum alloyed with some ten per cent rhodium than with pure platinum, but attempts over the years to eliminate the effect have been quite unsuccessful. Instead various filtration and gettering devices are employed to catch and recover the metal lost from the gauze during use.

The mechanism of the reaction is uncertain, the conditions not lending themselves to kinetic or mechanistic studies. It is distinctly possible that radical reactions initiated at the surface continue in the gas phase within the gauze bed.

This selective oxidation is uniquely performed by platinum alloys capable of withstanding the vigorous oxidizing conditions employed. Base metal oxides also catalyse ammonia oxidation, but the reaction is non-selective and the product is nitrogen.

The elimination of nitrogen oxides not absorbed in the scrubbing towers is discussed in the following chapter.

10.4. The oxidation of sulphur dioxide

The oxidation of sulphur dioxide to the trioxide is an essential step in the manufacture of sulphuric acid. Supported platinum was formerly used as catalyst, but it proved extremely susceptible to poisons such as arsenic and halogens. The platinum-catalysed reaction has been studied by several groups of workers, and kinetic equations have been presented. The process is exothermic, and temperatures of about 670 K give good equilibrium conversions. The catalyst now almost universally used is a supported molten salt formed from divanadium pentoxide and an alkali metal oxide, e.g. sodium vanadate. This is an example of a quite rare situation where the catalyst is liquid under operating conditions. It seems likely that lattice oxide is used for the oxidation, and the mobility of the catalyst ensures that a new and oxygen-rich surface is continually exposed to the reactants.

11. Catalysis in atmospheric pollution control

11.1. The role of catalysis in atmospheric pollution control

PUBLIC concern with environmental pollution has grown phenomenally in recent years. Although this concern is greatly to be welcomed, much of what has been written on the subject is not as scientifically accurate and objective as might be wished. The fact is that the exact risk to human beings, animals, and plants is not readily assessed. We know of course that high concentrations of reactive gases such as carbon monoxide, nitric oxide, and sulphur dioxide can be lethal. What is much more uncertain is their hazard at very low concentrations if inhaled over prolonged periods. Although we sometimes speak of a *safe threshold limit*, meaning the maximum safe concentration, this can only apply to the average person, or animal or plant. In any population, some of its members are more susceptible to the toxic effects of a pollutant than others: particularly at risk are young children, the elderly, and the chronically sick. We must note too that most of the impurities in the air arise from natural sources, which are not controllable: thus for example it has been estimated that no less than three-quarters of the sulphur dioxide in the atmosphere arises naturally. While it is wishful thinking to speak of the complete purification of our air, we have to be especially concerned with pollutant levels in places where high concentrations can accumulate, for example in factories and in city centres.

Atmospheric pollution in Great Britain reached a peak in the early 1950s. A particularly severe and protracted smog over London in 1952 resulted in some 4000 people dying prematurely through effects directly attributable to it. The most important cause of smog was the domestic coal-burning fire, which produces smoke and sulphur dioxide as by-products: this combination is especially harmful to persons suffering from bronchitis and other respiratory ailments. As a consequence of this disaster, the first Clean Air Act became law in this country in 1956. This contained the concept of smokeless zones, and their gradual introduction has led to a great improvement in the winter air quality in our major cities.

Unfortunately the decrease in the use of coal as a domestic fuel has been paralleled by a rapid growth in the number of cars and diesel-powered vehicles in and around our cities. As the figures in Table 11.1 show, cars were in 1968 the predonimant source of air pollution in the United States: recognition of this has led to plans (expected to become law in 1976 and 1977) for imposing very strict limits on exhaust emissions.

Table 11.1 lists the main classes of atmospheric pollutants, and a glance at the column headings shows which may be susceptible to control by catalytic

TABLE 11.1.

Air pollutant emissions in the United States in 1968 (in millions of metric tons)

Source	CO	SO_x	NO_x	Hydrocarbons	Particulate matter
Transportation	58·1	0·7	7·3	15·1	1·1
Electric power generation	0·6	15·2	3·6	0·2	5·1
Industry	9·1	11·2	4·8	4·3	9·3
Solid waste disposal	7·1	0·1	0·5	1·5	1·0
Miscellaneous	16·1	3·0	2·5	8·0	9·2
Totals	91·0	30·2	18·7	29·1	25·7

From *Nationwide Inventory of Air Pollutant Emissions 1968*, N.A.P.C.A. Publication No. AP-73 (U.S. Department of Health, Education and Welfare, Washington, D.C. 1970).

means. Quite clearly not much can be done with inorganic particulate matter (e.g. dust from power stations and cement works); neither can smoke and soot be readily treated by catalytic oxidation. Hydrocarbons and other organic substances can be catalytically burnt, and so of course can carbon monoxide. The nitrogen oxides, mainly nitric oxide and nitrogen dioxide, and symbolized as NO_x, present a quite different problem which is discussed further below. They have in fact to be reduced to nitrogen to render them harmless, and this reduction can be catalysed. Sulphur dioxide and trioxide, or SO_x, present even greater difficulties: emission of these substances from small power installations such as domestic oil-fired boilers could be reduced to elemental sulphur, but the resulting dust might become an even worse pollutant. Treatment of SO_x only becomes practicable at power stations where suitable means for recovery of the sulphur can be economically installed.

We therefore see that one of the principal areas in which catalysis can be employed in atmospheric pollution control is in oxidation of hydrocarbons, other organic substances, and carbon monoxide. This category can be further broken down in the following way.

(1) Abatement of industrial odours (see section 11.2).
(2) Treatment of vehicle exhaust emissions (see section 11.4).
(3) Purification of air in domestic and working environments.

Little attention has been paid to this last area, but it represents a large and profitable field for future exploitation.

The other chief application of catalysis is in the reduction of NO_x to nitrogen. The main, but not the only, process where this is a problem is the manufacture of nitric acid (see section 11.3). The oxidation and reduction uses conjoin in an interesting manner in the treatment of vehicle exhaust (see section 11.4). For further detailş of these and related problems, see D. J. Spedding *Air pollution* (OCS 20).

11.2. Abatement of industrial odours

Many industrial operations produce gaseous effluents containing traces of organic substances which smell horribly but which are not necessarily toxic. Such malodorous effluents may be regarded as polluting the atmosphere, and therefore need to be eliminated or abated. A partial list of processes producing unpleasant odours is the following: fat rendering, glue and size manufacture, tanning, fish-meal processing, manufacture and cutting of polyvinylchloride and polyurethane, food manufacture, coffee roasting, and manure processing.

There are basically three ways of abating these odours.

(1) Wet scrubbing of the gas, to dissolve the odiferous components in water.
(2) Absorption by activated carbon.
(3) Incineration, where fuel is mixed with the impure air and then burnt in a flame at about 1000 K.
(4) Catalytic combustion.

Each method has its own advantages and disadvantages. Methods 1 and 2 create subsequent effluent disposal problems; installation costs for method 3 are quite low but fuel costs are rather high; method 4 turns out to be cheaper than method 3 provided the catalyst survives more than about three months.

The catalytic combustion method is now beginning to be applied in a range of industrial problems. The preferred form of catalyst is the monolithic structure described in Chapter 4, with a thin coating of platinum deposited on it. This permits use of high space velocities with minimal pumping effort. One of the problems of catalytic combustion is how to heat the catalyst to the temperature (420–670 K) at which rapid oxidation of the pollutant will occur. The concentration of the pollutant is rarely such that its heat of combustion will maintain a high enough temperature. The best procedure is to heat the impure air to the desired temperature by means of a flame within the equipment, and then to use the hot exhaust gas to warm incoming air. Figure 11.1 shows schematically how this can be done. Catalyst poisoning can also be a difficulty. Phosphorus compounds are notorious poisons, and so catalytic combustion is not a viable proposition for gas streams containing them: thus air used for drying certain surface coatings (e.g. on car bodies) cannot always be treated in this way. Nevertheless the application of this method is certain to grow.

11.3. NO_x Abatement

The absorption efficiencies of scrubbers on nitric-acid plants lie between 98·2 and 99·3%: it is uneconomic to make them larger and hence more efficient. The effluent gas from a typical plant of 350 tons per day capacity has a volume of $3{\cdot}4 \times 10^3$ m^3 hr^{-1}, and contains 0·08 to 0·30% NO_x, 2 to 3%

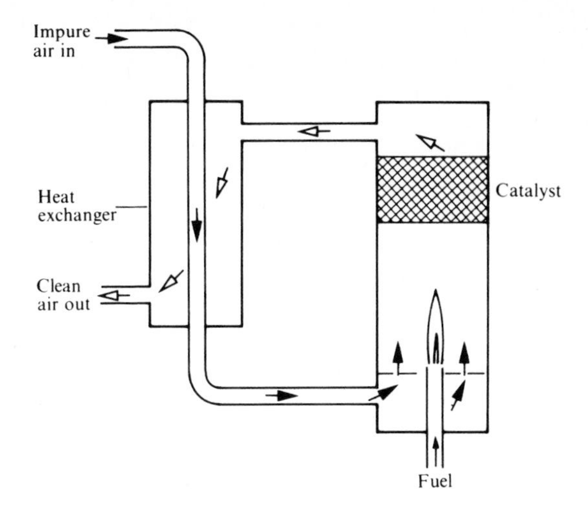

FIG. 11.1. Diagram of equipment for catalytic odour abatement.

oxygen, the balance being nitrogen and water vapour. Several different fuels may be used to reduce the NO_x to nitrogen: these include hydrogen, methane, propane, or butane. The first group of reactions which occurs when the effluent containing added fuel is passed over a catalyst are called *decolorization reactions*. In these the oxygen is reduced to water and nitrogen dioxide to nitric oxide, e.g.

$$H_2 + NO_2 \rightarrow H_2O + NO.$$

Since nitric oxide is colourless, the effluent after such a treatment would appear to be innocuous, but since nitric oxide reoxidizes to nitrogen dioxide in the atmosphere the problem has not really been cured. It is therefore necessary to achieve, by the use of higher concentrations of fuel, the *abatement reaction*, e.g.

$$2H_2 + 2NO \rightarrow 2H_2O + N_2.$$

It is usually thought necessary to reduce the concentration of NO_x to less than 200 parts per million (p.p.m.).

In practice the effluent from the scrubber is heated by heat exchange with the hot gas emerging from the reactor. The minimum inlet temperature for high conversion depends upon the fuel used (hydrogen, 470 K; propane and butane, 520 K; methane 750 K). A great deal of heat is also liberated by the combustion of the fuel, and this has to be used (by means of a turbine) to generate energy for the process to be economic. For every per cent of oxygen in the gas stream there is a temperature rise of about 160 K when hydrogen is the fuel, or about 130 K when a hydrocarbon fuel is used. This effect puts a limit on the amount of oxygen that can be removed in a single stage.

One of the most successful types of catalyst used for NO_x abatement is a platinum-coated ceramic honeycomb of the kind described in section 4.3. A monolithic structure is particularly useful in this application because of its strength, thermal stability, and low resistance to gas flow.

11.4. The control of motor-vehicle exhaust gases

Table 11.1 has already indicated the importance of the internal combustion engine (ICE) as an atmospheric polluter. The magnitude of the problem is further stressed by the following conservatively estimated annual emissions from a new American 1963 model car: 750 kg of carbon monoxide, 250 kg of hydrocarbons, and 40 kg of nitrogen oxides, these being formed from atmospheric nitrogen and oxygen in the combustion stage. While the carbon monoxide and nitrogen oxides come entirely from the exhaust gases, only 65 per cent of the hydrocarbons arise in this way: the remainder stem from the crankcase blowby (20%), the carburettor (9%) and the fuel tank (6%). Diesel engines are comparatively minor sources of pollution.

Awareness of the hazards of pollution from the ICE arose through the gradual recognition of their responsibility for 'photochemical smog', first detected in the Los Angeles area in the early 1940s. Because the city is enclosed on three sides by high ground, and since vertical air circulation is often restricted because of a temperature inversion, it is particularly prone to the phenomenon; but it has since been detected in Tokyo and some other large cities, although it is fortunately not a problem in Britain. Photochemical smog is formed by a complex series of reactions between hydrocarbons, nitrogen oxides, and oxygen catalysed by sunlight. It affects plant growth, and at low concentrations (~0·1 p.p.m.) it is irritating to the eyes, while at higher concentrations (>0·6 p.p.m.) it affects pulmonary functions. The principal noxious agent is peroxoacetyl nitrate:

$$CH_3—\overset{\overset{\displaystyle O}{\|}}{C}—O—O—NO_2$$

For further details see Spedding (OCS 20).

This situation has led to the introduction, first in California and later extending to the whole of the United States, of legislation to limit ICE emissions. Many solutions to the problem have been proposed. We cannot concern ourselves here with radical solutions like changing to a cleaner fuel (e.g. propane or methanol) or complete redesign of engine operation (e.g. the Wankel engine): nor should we be interested in purely thermal reactors placed behind the exhaust manifold, wherein it is hoped the combustion will be completed before the gases cool too much. Our attention must be limited to the use of catalysts to finish the oxidation, and to remove the nitrogen oxides.

Let it be said at once that the problem of using catalysts for this purpose is a most difficult and challenging one. The following are its principal features.

(1) It is required to remove the NO_x by its reaction with carbon monoxide or unburned hydrocarbons by a reaction such as

$$2NO + 2CO \rightarrow N_2 + 2CO_2.$$

There is adequate reductant except when the air–fuel ratio is just on the lean side of stoichiometric, which is about 15:1 (see Fig. 11.2). It is important that reduction does not stop at the formation of ammonia, since this might later reoxidize to NO_x.

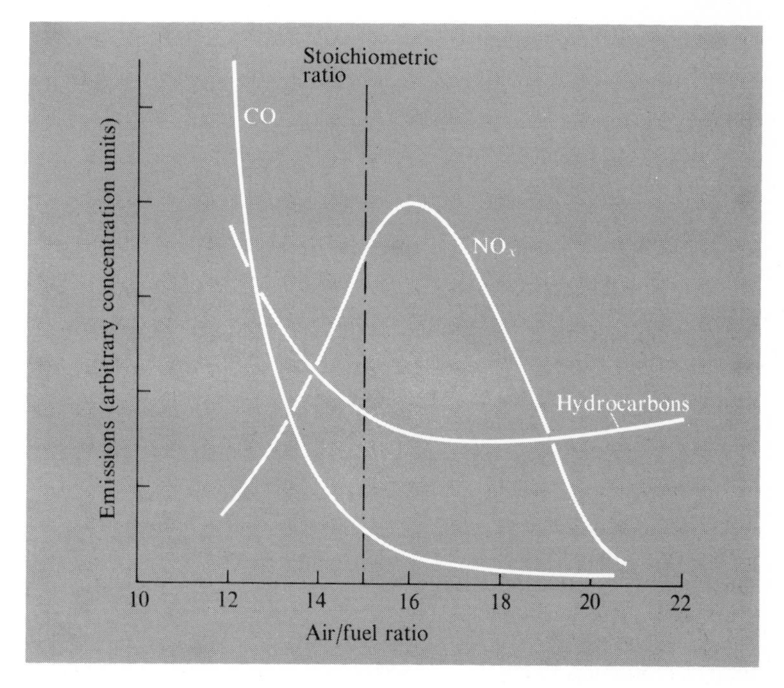

FIG. 11.2. Dependence of ICE exhaust emissions upon air/fuel ratio.

(2) Excess carbon monoxide and unburned fuel must be oxidized to carbon dioxide and water.

(3) The catalyst should start to operate at the lowest possible temperature, since emissions are high during the first few seconds of engine running, and should also be effective at normal operating temperatures (570 K to 770 K, depending on distance of the catalyst from the manifold) and be

able to withstand for short periods the much higher temperatures (around 1270 K) that might occur during engine malfunction.

(4) The catalyst must work properly for the time taken for the vehicle to travel 50 000 miles or for five years, whichever is the shorter time.

(5) It must be active enough to produce the desired conversions at space velocities up to 150 000 hour^{-1}.

The greatest stumbling block to the use of catalysts is undoubtedly the fact that they are rapidly poisoned by the lead compounds normally present in petrol as anti-knock agents and the phosphorus compounds present for pre-ignition control: catalysts based on the noble metals are especially susceptible to such poisoning. In anticipation of the ultimate introduction of catalytic afterburners, several oil companies are already offering 'lead-free' petrol in the United States: its effective use will however entail a decrease in the compression ratio of car engines.

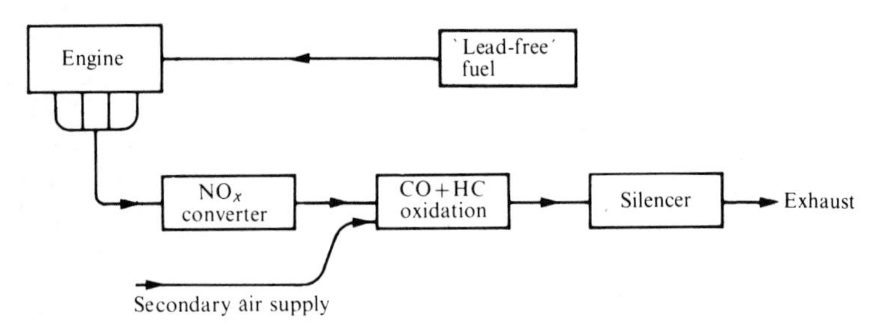

FIG. 11.3. Diagrammatic representation of a catalytic system for treating ICE emissions.

The system which is envisaged is shown diagrammatically in Fig. 11.3. The catalysts for both the NO_x converter and for the hydrocarbon and carbon monoxide oxidation seem likely to be promoted platinum composition supported on a monolithic support to allow the necessarily high space velocities without an undesirable pressure build-up. While base metal formulations are effective above about 600 K, they are ineffective in the first critical seconds of running. Catalyst employed in non-selective oxidations or other highly exothermic reactions exhibit a phenomenon known as *light-off temperature*. This is the temperature at which the heat of reaction is produced rapidly enough to raise the temperature of the catalyst well above that of its surroundings, with a consequent very rapid increase in rate of conversion. Promoted platinum catalysts have been developed with a light-off temperature for hydrocarbons and carbon monoxide as low as 440 K.

It is uncertain at the time of writing whether any kinds of catalysts will be used commercially for the control of vehicle exhaust emissions; if they are, it will be the most demanding application that catalyst designers have ever had to work on.

Bibliography

NOTE ON THE LITERATURE OF CATALYSIS

UNTIL about 1960, original papers and review articles on catalysis were widely scattered throughout the chemical literature, but while this is still partly true there are now a number of journals and books to which the interested reader should first direct his attention. Original papers are to be found in the *Journal of Catalysis* (Academic Press) which started in 1962, and in the Russian journal *Kinetika i Kataliz* which also appeared at about this time. *Advances in Catalysis* (Academic Press), which contains review articles, was first published in 1948, and *Catalysis Reviews* (Dekker), also a review journal, started in 1968. There has been a series of International Congresses on Catalysis, the proceedings of which have been published: the details are as follows.

Number	*Year*	*Location*	*Publisher*
First	1956	Philadelphia	Academic Press (as vol. 9 of *Advances in Catalysis*)
Second	1960	Paris	Editions Technip
Third	1964	Amsterdam	North Holland
Fourth	1968	Moscow	Akadémiai Kiadó, Budapest
Fifth	1972	Palm Beach, Fla.	North Holland

Each contains plenary lectures, original papers and the ensuing discussion. Several Discussions of the Faraday Society (now the Faraday Division of the Chemical Society) have been devoted to adsorption and catalysis (e.g. in 1950 and 1966). Textbooks and monographs are mentioned in the following paragraphs.

GENERAL TEXTBOOKS, MONOGRAPHS, AND REVIEW ARTICLES

Publications listed in this section relate particularly to Chapters 1–4 of this book; they are termed either elementary or advanced, these terms being of course merely relative.

ELEMENTARY

ROBERTSON, A. J. B. *Catalysis in chemistry*, Methuen Educational Ltd., London (1972). A useful introduction to academic aspects of catalysis with interesting historical background.

THOMSON, S. J. *and* WEBB, G. *Heterogeneous catalysis*, Oliver and Boyd, Edinburgh (1968). Selective in the treatment of the subject, with some rather advanced sections.

BOND, G. C. Catalysis in the context of chemistry. *RIC Reviews*, **3**(**1**), 1 (1970). Emphasizes solid-state aspects of catalysis, and discusses physical and chemical approaches to understanding them.

WEBB, G. Some aspects of catalysis by solid surfaces, *Science Progress*, **60**, 337 (1972). A short general review.

BOND, G. C. *Principles of catalysis* (2nd edn, revised), Chemical Society, London (1972). Concentrates on basic principles, the level of treatment being about the same as in this book.

PRETTRE, M. (trans. D. ANTIN). *Catalysis and catalysts*, Dover Publications, New York (1963). Now rather out of date, but still a valuable general survey.

ADVANCED

THOMAS, J. M. *and* THOMAS, W. J. *Introduction to the principles of heterogeneous catalysis*, Academic Press, London and New York (1967). More than just an introduction: a comprehensive treatment, with particular emphasis on heat and mass transfer.

RIDEAL, E. K. *Concepts in catalysis*, Academic Press, London and New York (1968). A selective and highly personal view of problems in catalysis by the grand old man of the subject, writing with half a century's experience in the field.

ROBERTSON, A. J. B. *Catalysis of gas reactions by metals*, Logos Press, London (1970). An academic treatment of reactions on clean metal surfaces.

HAYWARD, D. O. *and* TRAPNELL, B. M. W. *Chemisorption* (2nd edition), Butterworths, London (1964). A full description of chemisorption on metals and oxides, now unfortunately somewhat dated.

CLARK, A. *The theory of adsorption and catalysis*, Academic Press, London and New York (1970). A useful survey of theoretical concepts.

ANDERSON, J. R. (editor). *Chemisorption and reactions on metallic films*, vols. 1 and 2, Academic Press, London and New York (1971). Another academic treatment of adsorption and reactions on clean metal films, but much more detailed than Robertson's book.

BOND, G. C. *Catalysis by metals*, Academic Press, London and New York (1962). A monograph on adsorption and catalysis written principally from an academic standpoint, now rather out of date.

KRYLOV, O. V. (trans. M. F. DELLEO, J. HAPPEL, G. DEMBINSKI, and A. H. WEISS), *Catalysis by non-metals*, Academic Press, London and New York (1970). Useful because of its especially full treatment of Russian work, but unfortunately omitting much recent work on mixed oxide systems.

GREGG, S. J. *and* SING, K. S. W. *Adsorption, surface area, and porosity*, Academic Press, London and New York (1967). A full and critical presentation of concepts relevant to physical adsorption of gases on porous solids.

LINSEN, B. G. (editor). *Physical and chemical aspects of adsorbents and catalysts*, Academic Press, London and New York (1970). An up-to-date discussion of a number of specialized topics in the field of catalysts and catalysis.

BASOLO, F. *and* BURWELL, R. L. *Catalysis—progress in research*, Plenum Press, London and New York (1973). A *tour de l'horizon* of modern research in all areas of catalysis, including enzymatic catalysis, with suggestions of promising areas for further work.

SOKOL'SKII, D. V. *Hydrogenation in solutions*, Israel Program for Scientific Translations, Jerusalem (1964). Not wholly confined to liquid phase reactions, but valuable for its discussion of the Russian literature.

SPECIFIC REFERENCES

This section contains references pertinent to Chapters 5–11.

Chapter 5

DUTTON, H. J. Some new approaches in lipid research. *Chem. and Ind.*, 665 (1972).

ALBRIGHT, L. F. Commercial processes for hydrogenating fatty oils. *Chemical Engineering*, 249 (1967).

COENEN, J. W. E. Hydrogenation of oils and fats. *J. Oil Technologists Association of India*. 16 (1969).

WELLS, P. B. Some aspects of the selective action of metal catalysts. *Surface and defect properties of solids*, **1**, 236. The Chemical Society, London (1972).

BOND, G. C. *and* WELLS, P. B. The mechanism of the hydrogenation of unsaturated hydrocarbons on transition-metal catalysts. *Adv. Catalysis*, **15**, 92 (1964).

Chapter 6

FREIFELDER, M. *Practical catalytic hydrogenation*, Wiley-Interscience, New York (1971).

RYLANDER, P. N. *Catalytic hydrogenation over platinum metals*, Academic Press, London and New York (1967).

AUGUSTINE, R. L. *Catalytic hydrogenation*, Dekker, New York (1965).

Chapter 7

TANAKA, K. *Solid acids and bases*, Academic Press, London and New York (1970).

GERMAIN, J. E. *Catalytic conversions of hydrocarbons*, Academic Press, London and New York (1969).

Chapter 8

STERN, J. P. *and* STERN, E. S. *Petrochemicals today*, Arnold, London (1971).

SAMUEL, D. M. *Industrial chemistry—organic* (2nd edn), Royal Institute of Chemistry, London (1972).

STANLEY, H. M. *The petroleum-chemicals industry*. Royal Institute of Chemistry, London (1963).

VOGE, H. H. *and* ADAMS, C. R. Catalytic oxidation of olefins. *Adv. Catalysis*, **17**, 151 (1967).

Chapter 9

DOWDEN, D. A. and others, *Catalyst handbook*, Wolfe Scientific Books, London (1970).

Chapter 10

VANCINI, C. A. (trans. L. PITT), *Synthesis of ammonia*, Macmillan, London (1971).

DOWDEN, D. A. and others. *Catalyst handbook*, Wolfe Scientific Books, London (1970).

Chapter 11

SEARLES, R. A. Removal of odours by catalytic incineration. *Environmental Pollution Management*, **2**, 280 (1972).

HODGES, L. *Environmental pollution*, Holt, Rinehart, and Winston, New York (1973).

ACRES, G. J. K. *and* COOPER, B. J. Automobile emission control systems, *Platinum Metals Review*, **16**, 74 (1972).

OXFORD CHEMISTRY SERIES

A number of references have been given in this book to other volumes in the series. A complete list appears at the front of the book, and those that are closely connected with this book are listed here.

PUDDEPHATT, R. J. *The periodic table of the elements* (1972).

SMITH, E. B. *Basic chemical thermodynamics* (1973).

COULSON, C. A. *The shape and structure of molecules* (1973).

WORMALD, J. *Diffraction methods* (1974).

STERN, E. S. (editor). *The chemist in industry* (1): *fine chemicals for polymers* (1973).

EARNSHAW, A. and HARRINGTON, T. J. *The chemistry of the transition elements* (1973).

FYFE, W. S. *Geochemistry* (1974).

STERN, E. S. (editor). *The chemist in industry* (2): *human health and plant protection* (1974).

SPEDDING, D. J. *Air pollution* (1974).

PILLING, M. J. *Reaction kinetics* (1974).

Index

Catalysts

MoO_3 p 35

Sb_2O_5 p 35

V_2O_5 p 35

ZnO p 35